Dedicated to the Memory

of

Gloria 'Timmy' Wilson

SPITFIRE

Stewart Wilson

INTRODUCTION

"Oh no, not another Spitfire book!", I hear you say. Yes it is, but one which attempts to present a history of arguably the most famous fighter in a slightly different way.

The book's several major sections comprise a potted narrative history of the aircraft; a detailed review of the various Spitfire and Seafire marks and models (with specification and performance data); a summary of foreign operators; squadron summary; a comprehensive serial number table; and a chronological section which tells the Spitfire/Seafire story in 'diary' form.

The idea with this different approach is to put what is a complicated tale into some kind of perspective. There was so much going on with the development of the Spitfire – with various marks and models simultaneously being developed, built and put into service – it's easy to lose track of things and more importantly, to lose the appreciation of how much work was performed in a very short time and how much pressure those involved were under.

I make no claim to the chronology section being definitive but hope it does provide a different perspective to the Spitfire/Seafire story, as well as introducing a few snippets of technical, personal and operational information which may otherwise be missed. Included in the chronological section are numerous breakout boxes covering various topics associated with the Spitfire, these designed to fill in a few more aspects of what is a fascinating story.

I hope readers find something of interest within these pages.

My thanks to Jim Thorn and all at Aerospace Publications for their ongoing enthusiasm and support and especially to production manager Gayla Wilson, Scott Mason for sourcing most of the photographs and to him and Gerard Frawley for looking after the proofreading. Thanks also to Juanita Franzi for her drawings and to Neil Mackenzie and Mike Kerr for supplying photographs from their collections.

Finally and most importantly, my thanks, love and gratitude to Wendy, without whom it would have been impossible to get through a period involving both a horrendously heavy workload and a family tragedy.

Stewart Wilson
Buckingham 1999

Published by Aerospace Publications Pty Ltd (ACN: 001 570 458) PO Box 1777, Fyshwick, ACT 2609, Australia.
Phone (02) 6280 0111, fax (02) 6280 0007, e-mail mail@ausaviation.com.au and website
www.ausaviation.com.au – publishers of monthly *Australian Aviation* magazine.
Production Manager: Gayla Wilson

ISBN 1 875671 45 5

Proudly Printed in Australia by Pirie Printers Pty Ltd, 140 Gladstone St, Fyshwick, ACT 2609.
Distributed throughout Australia by Network Distribution Company, 54 Park St, Sydney, 2000. Fax (02) 9264 3278
Distributed in North America by Motorbooks International, 729 Prospect Ave, Osceola, Wisconsin, 54020, USA.
Fax (715) 294 4448. Distributed throughout Europe and the UK by Airlife Publishing Ltd, 101 Longden Rd, Shrewsbury SY3 9EB, Shropshire, UK. Fax (743) 232944.

FRONT COVER: Spitfire prototype K5054; Spitfire IA R6800 of 19 Squadron RAF; Spitfire HF.VIII A58-614 of 457 Squadron RAAF; Spitfire F.VB EP837 of 308th FS, 31st FG USAAF; Seafire FR.47 VP461 of 800 Squadron RN FAA. (artwork by Juanita Franzi)

CONTENTS

Although seen many times before, Charles Brown's famous shot of Spitfire XIV RB140 is worth repeating because it sums up the aircraft perfectly: power, elegance, brute force and grace in a single package.

SPITFIRE SUPERB

SPITFIRE SUPERB

Mention the word 'Spitfire' to the world's aviation enthusiasts and the superlatives will inevitably flow from many of them – 'the world's greatest fighter', 'World War II's greatest fighter', 'the world's most famous fighter', and so on. The Spitfire has certainly acquired a mystique of legendary proportions in the 60 plus years since its first flight and entry to service, becoming a symbol of fighting the good fight and resisting the intentions of the invader.

Legend is one thing; usefulness as a fighting machine is another, and it is in this context that Reginald Mitchell's sleek single engined fighter design must be considered. The description 'greatest fighter' is simply not appropriate when applied to the Spitfire or any other aircraft for that matter. 'One of the greatest' would certainly be nearer the case, and in the context of World War II, one of top three (along with the North American P-51 Mustang and Focke-Wulf Fw 190) seems to be the level generally agreed upon by knowledgeable enthusiasts.

The Spitfire and the Mustang are obvious aircraft to compare as when it comes to Allied fighter aircraft of the period, these are the two which

most enthusiasts get excited about. But caution is required as a closer look reveals two outwardly similar (single engined, single seat, Rolls-Royce powered) aircraft which are in fact quite different in capability.

The Mustang cannot be touched when it comes to versatility of role and the important factor of long range. But as a dogfighter, a late model Spitfire will always beat a Mustang given equal pilot skills. Like most other British fighter aircraft since, the Spitfire's relatively meagre internal fuel capacity proved to be a handicap for missions other than point defence, but its true worth remains that in its extensively developed versions it was – apart from a brief period in 1941/42 – always able to at least hold its own as an interceptor fighter during the course of World War II.

At one stage it became fashionable to attempt to decry the value of the Spitfire to the Allies' and Britain's war effort.... an example of the 'tall poppy syndrome' at work if you like. The Battle of Britain in 1940 was when the Spitfire legend developed, where the aircraft became a symbol freedom and where myths surrounding its charismatic good looks and

flying qualities were born. The Spitfire was, as a result, credited with rather more than it possibly deserved during that immensely important battle and has always tended to overshadow its less photogenic partner, the Hawker Hurricane.

Simple logic tells us that as there were more Hurricanes on Royal Air Force strength in 1940, that aircraft must have borne the brunt of the fight. But neither aircraft could have thwarted the *Luftwaffe* on its own.

Two simple facts tend to reinforce the Spitfire's worth; 20,351 were built before, during and after the war (plus 2,408 navalised variants, the Seafire) and the aircraft was the only Allied fighter to be in production when war broke and still be coming off the line six years later when it was all over. It is difficult to imagine the users of the Spitfire allowing those milestones to be achieved if the product they were buying was not up to scratch. The fact that the Spitfire was able to maintain its competitiveness against much later designs speaks for itself.

The secret to the Spitfire's success lay in the ability of the basic design to be continuously modified and upgraded as the need arose. In fact, there was a case for the later models to acquire a new name as they bore little resemblance to the early versions, perhaps in the Rolls-Royce Griffon engined variants and almost certainly in the final marks when an entirely new wing was developed. However, the same argument could be applied to the Mustang when it changed from Allison to Rolls-Royce power and underwent a redesign of its fuselage.

If the Spitfire Mk.I of 1938 and the Mk.24 of 1946 are compared it can be seen that the ultimate variant had a different engine, fuselage, wing, cockpit canopy, tail surfaces, undercarriage and armament! Its maximum takeoff weight, available horsepower, wing loading and rate of climb all approximately doubled, speed increased by around 100mph (161km/h), weight of fire from the internal guns trebled, service ceiling increased by 10,000 feet (3,050m), the rate of roll nearly quintupled and the permissible centre of gravity range more than quadrupled.

Integral with these quite astounding statistics – particularly given the short time span involved – is the parallel development of the two vee-configuration, 12-cylinder, liquid-cooled engines provided by Rolls-Royce to power the Spitfire.

The 27-litre Merlin's legendary status almost rivals that of the Spitfire and it's worth remembering that if the powerplant had not been developed to the extent it was, the Spitfire (and other aircraft including the

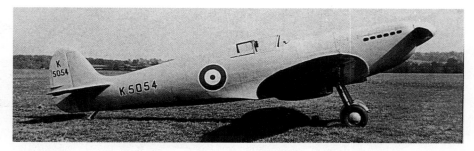

The first and the last: the prototype Spitfire (top) and the Seafire FR.47, ten years and a world apart in terms of performance, mechanical specification and firepower.

Mustang) would not have achieved what it did.

The second engine used in the Spitfire, the Griffon, was of similar configuration to its predecessor but of larger 37-litre cubic capacity and was thus able to produce more power than the Merlin while using the same supercharger technology rapidly developed for the earlier engine. Early Merlins used single-stage/single-speed superchargers which operated most efficiently at medium altitudes, but by 1942 two-speed/two-stage blowers were in front line use, with the optimum height at which maximum power was produced able to be 'tailored' by changing the supercharger gear ratios.

Some more statistics emphasise the Spitfire's importance: 29 major marks and 59 major sub-types were produced (including the Seafire), the Spitfire served with 179 Royal Air Force operational squadrons between 1938 and 1954 and the Seafire with 25 Fleet Air Arm squadrons 1942-50; and wartime users apart from Britain included Turkey, the Free French, Portugal, Italy, Yugoslavia, Egypt, the United States, the Soviet Union and Australia.

The Art of High Speed Flight

By the mid 1930s, the small Supermarine company and its chief designer, Reginald Mitchell, were in something of a unique position among aircraft manufacturers. Its product range to date had consisted of two remarkably dissimilar ends of the aviation scale – on one hand the stately biplane flying boats and amphibians (Scapa, Stranraer, Seagull,

Some of the Supermarine family in 1936, the photograph taken shortly after the prototype Spitfire's maiden flight. Left to right: 'Mutt' Summers (chief test pilot), H J Payne (technical assistant to R J Mitchell), Reginald Mitchell (chief designer), Stuart Scott-Hall (from the Air Ministry) and Jeffrey Quill (assistant to Summers).

The Southampton flying boat, Supermarine's most numerous product before the Spitfire arrived. By 1936, Supermarine had manufactured a grand total of only 137 aircraft.

Southampton) hardly represented the leading edge of high speed flight.

On the other hand, Mitchell's series of very high speed monoplane floatplanes certainly did, and in 1931 one of them – the S.6B – set a new world air speed record of 407mph (655km/h). The greatness of this achievement is well illustrated by the fact that it would be more than a decade before operational fighter aircraft from any country could routinely achieve that kind of speed.

An international contest called the Schneider Trophy was the catalyst for the S.6B and predecessors. Established by Frenchman Jacques Schneider in 1912, the idea of the contest was to promote the development of aviation – particularly that of water-borne aircraft – and the rules were that contestants had to be such an aircraft, that a race would be held over a set course and that if any nation won the contest on three successive occasions it would become the permanent holder of the trophy.

Between the first contest in 1912 and the last in 1931 the speeds of the aircraft increased dramatically, as did the technology used in the airframes and powerplants. Reginald Mitchell's and Supermarine's first contribution to Britain's effort came in 1922 when the Sea Lion II biplane won the event at an average speed of 145.7mph (234.4km/h).

After that, the monoplane reigned supreme, Mitchell's wooden S.4 floatplane setting a new world air speed record of 226mph (364km/h) in 1925 but failing to contest the Schneider due to a crash in trials.

In 1927 the all metal S.5 (900hp/671kW Napier Lion engine) won at a speed of 281.6mph (453.3km/h), followed two years later by the slightly larger S.6 with its vee-12 Rolls-Royce R engine tuned to produce 1,900hp (1,417kW). The 'R' would later form the basis of the production Griffon.

Britain laid permanent claim to the Schneider Trophy in 1931 when the Supermarine S.6B made it three in a row. This developed version of the original S.6 featured a 2,350hp (1,752kW) version of the Rolls-Royce R and recorded a speed of 340.1mph (547.3km/h) over the course at Calshot in England. This was over 100mph (161km/h) faster than any front line fighter of the time was able to achieve.

It would be wrong to suggest that these high speed floatplanes had any direct bearing on the design of the Spitfire, but if nothing else they left Mitchell and his team with a thorough knowledge of the aerodynamic and structural considerations when designing such an aircraft, all of which had obvious application to fighter design. Similarly, Rolls-Royce's experiences with the 'R' engine for the S.6 and S.6B – particularly in the areas of supercharging, fuel concoctions and the stresses involved with high revs and power outputs – would stand it in good stead when Supermarine and other airframe manufacturers began calling for ever more power to keep their aircraft competitive in war.

The Type 224

Set against this background, it's surprising that Mitchell's and Supermarine's first attempt at a fighter was disappointing in performance. The Supermarine Type 224 was designed to meet the requirements of the RAF's specification F.7/30, issued to find a replacement for the ageing Bristol Bulldog biplane fighter.

Supermarine's submission revealed a largish cantilever 'inverted gull' low wing monoplane (considered advanced at the time) with a 'trousered' fixed undercarriage, open cockpit and 600hp (447kW) Rolls-Royce Goshawk evaporative cooled engine. As per the specification, the 224 was armed with four 0.303in machine guns. It made its first flight in February 1934 and immediately proved disappointing, exhibiting a top speed of only 238mph (383km/h) and a sluggish rate of climb.

There were also major problems with the complicated cooling system and the RAF contract went to one of the seven other designs submitted, the Gloster SS37. This went into production as the RAF's last biplane fighter, the Gladiator.

Towards Spitfire

Undeterred, Mitchell immediately set about improving the 224 and several designs evolved under the general heading of Type 300, the type number eventually bestowed upon the prototype Spitfire as it finally emerged.

The first revision involved lopping six feet (1.83m) off the 224's wing span and fitting retractable undercarriage and an enclosed cockpit. The Goshawk engine was retained but despite a projected top speed of nearly 270mph (435km/h) the Air Ministry was still unimpressed and it was back to the drawing board.

The next stages involved the fitting of a new, much thinner wing (still straight tapered in plan form) and Rolls-Royce's new 12-cylinder

The sole Supermarine Type 224 fighter. First flown in February 1934, the 224 lost the F.7/30 contract to the Gloster Gladiator, the last biplane to enter RAF service.

vee, liquid cooled engine, the PV.12. The 'PV' part of the designation stood for 'Private Venture', '12' for the number of cylinders. This engine would eventually become the Merlin and the first example had been bench tested in October 1933. By November 1934 the PV.12 had still not taken to the air but had completed its 100 hour Type Test and was rated at 790hp (589kW) at 12,000 feet using a supercharger boost pressure of +2lb. Eight years later, special Merlins would be producing over 2,000hp (1,491kW) running +25lb and using 150 octane fuel, while normal production examples would be capable of producing more than 1,700hp (1,277kW) at +18lb boost while using standard 100 octane fuel.

In November 1934, Supermarine submitted its revised PV.12 powered Type 300 design to the Air Ministry, which two months later wrote a specification around it (F.37/34) and provided £10,000 to cover the cost of building a prototype.

But that wasn't the end of the evolution of the Type 300, as the famous elliptical wing still had to come, as did the armament of eight 0.303in Browning machine guns. These evolved during continual design refinement during 1935, as did the other features which would result in the final prototype configuration.

Design For Performance

The elliptical wing provided an efficiently thin section yet also a sufficiently wide chord to house the outwards retracting undercarriage, the guns and other bits and pieces which had to be housed within it.

Despite the apparently conflicting requirements of light weight, fineness and strength, a very strong wing was designed for the Spitfire and this required a fairly complicated all metal internal structure comprising a single spar made up of spar booms joined by thick web plate; forward of that was the area providing most of the wing's strength, a very strong torsional box leading edge which could be easily further strengthened simply by substituting thicker gauge alloys. The wing was covered in Alclad and the control surfaces (including the rudder and elevators) were fabric covered.

This design was a bit of a two-edged sword in that while it provided for extensive future development,

Reginald Mitchell (left) and Sir Henry Royce in 1931 at the time of the S.6B's Schneider Trophy and World Air Speed Record triumphs.

aerodynamic efficiency and the ability to absorb high 'g' loadings, it would also prove somewhat troublesome to build in quantity, especially by the tens of thousands of previously unskilled hands which worked on the production lines during the war.

The man responsible for the structural design of the Spitfire's wing was Supermarine's chief draughtsman, Joe Smith. Following Reginald Mitchell's untimely death from cancer in 1937 at the age of 42, Smith would take over as chief designer and was thus the driving force behind the extraordinary development of the basic Spitfire design during the war.

The slender Spitfire fuselage was of all metal stressed skin monocoque construction with transverse frames, 'Z' section intercostal longitudinals, four main longerons and flush rivetted Alclad skinning. The tail unit was built integrally with the rear fuselage, fuel was housed in two tanks immediately behind the engine and in front of the pilot totalling 85 imperial gallons (386 litres) and the relatively narrow tracked outwards retracting main undercarriage was augmented by a fixed tail skid on the prototype and a wheel on production models.

Split flaps were fitted under each wing between the ailerons and the

Supermarine's first design to go into quantity production, the Walrus amphibian. First flown in June 1933 as the Seagull V for the Royal Australian Navy, production eventually reached 771 between then and early 1944. This RAN aircraft is being catapulted from the cruiser HMAS Australia. (RAN)

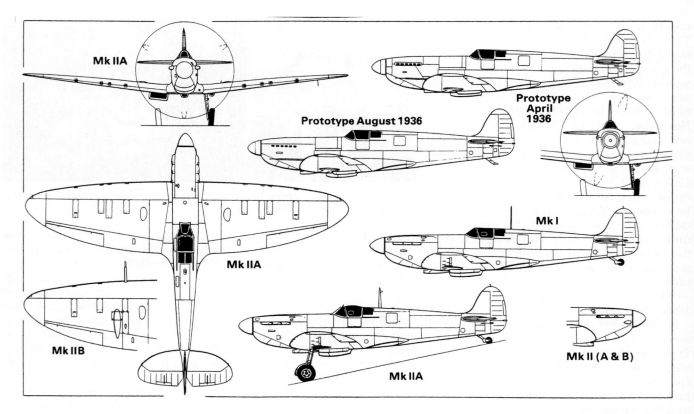

Mk IIA

Prototype August 1936

Prototype
April
1936

Mk I

Mk IIA

Mk IIA

Mk IIB

Mk II (A & B)

fuselage, these pneumatically operated (as were the guns and radiator shutters) and being either fully up or fully down. The 'down' position was initially 57 degrees; this was later changed to 90 degrees to cure a tendency to float just before touchdown. The hydraulic system operated the undercarriage only.

Like early production Spitfires, the prototype had an asymmetric appearance in end elevation due to the large engine cooling radiator under the starboard wing and the much smaller oil cooler under the port wing. Air for the engine's updraught powerplant came from a small intake under the nose in line with the wing leading edges.

The Spitfire Flies

Fitted with a 990hp (738kW) pre production Rolls-Royce Merlin C engine and a de Havilland two bladed fixed pitch propeller, the prototype Spitfire (K5054) made its first flight from Eastleigh Aerodrome on 5 March 1936 in the capable hands of the company's chief test pilot Captain J 'Mutt' Summers. For that flight a fine pitch propeller was fitted and the undercarriage locked down. Externally, the prototype differed from the early production examples by virtue of its tail skid and flush exhaust ports.

Initial flights revealed a disappointing maximum speed of only 335mph (539km/h), only a fraction faster than the less advanced but similarly powered Hawker Hurricane prototype had been achieving. Continual refinement which included the fitting of different propellers, a slightly more powerful Merlin F (later produced as the Merlin II) and the 'kidney' type ejector exhausts which would be fitted to production aircraft would soon have the speed up to a more respectable 350mph (563km/h).

By today's standards it seems remarkable that K5054 remained the only Spitfire flying for more than two years before the first production model took to the air in May 1938. As such, the prototype was responsible for all the development work during that period, and despite a couple of minor accidents on the way, it survived until 4 September 1939 when it was written off in a fatal landing accident at Farnborough. By this time the aircraft had been modified to almost full production Mk.I specifications and had flown somewhere in the region of 260 hours.

K5054 had cost the princely sum of £20,738 of which £12,478 had been borne by the British taxpayer, £7,500 by Rolls-Royce and the remainder by Supermarine. Supermarine's efforts were rewarded in June 1936 when

the Air Ministry ordered 310 production aircraft, following this with another order for a further 200 Spitfires in 1937. With war clouds looming over Europe in the late 1930s and the RAF's expansion schemes in place, additional orders had increased the total to 2,160 by the time Britain declared war on Germany in the first days of September 1939.

These orders created problems for Supermarine, whose previous production efforts had been in modest quantities. It was therefore decided early in the Spitfire's career that a 'shadow' production facility would be established at the Nuffield Organisation's Castle Bromwich factories in Birmingham, and this went on to become the major source of Spitfire production. Of the 22,759 Spitfires and Seafires ultimately built, just over half (11,989) emerged from what was called the Castle Bromwich Aircraft Factory (CBAF) from mid 1940. Westland and Cunliffe-Owen Aircraft also contributed to the tally (2,157 and 504, respectively) mainly by building Seafires.

Supermarine's own production was spread among several facilities. At first it was centred at its Woolston (Southampton) factory with main assembly and flight testing at Eastleigh Aerodrome. Expansion to a new site on reclaimed land on the River Itchen was soon in place but

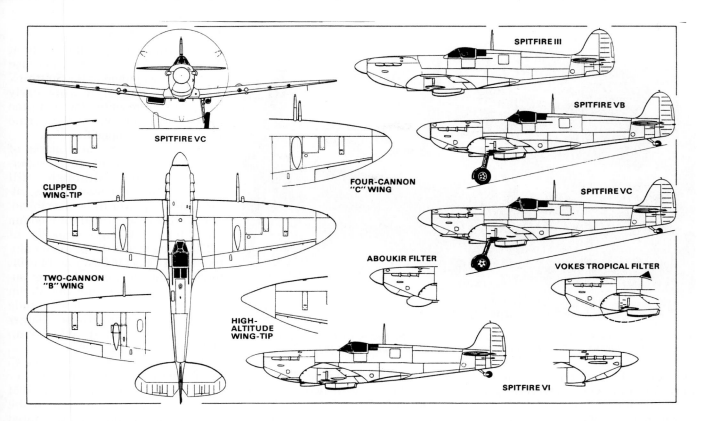

SPITFIRE III
SPITFIRE VB
SPITFIRE VC
SPITFIRE VC
SPITFIRE VI
CLIPPED WING-TIP
TWO-CANNON "B" WING
FOUR-CANNON "C" WING
HIGH-ALTITUDE WING-TIP
ABOUKIR FILTER
VOKES TROPICAL FILTER

manufacturing work was also placed with numerous garages and workshops in the Southampton area when the Battle of Britain started in anticipation of the inevitable raids on the main factories.

These started in August 1940 with little effect but a series of concentrated raids on the Itchen and Woolston facilities in late September did substantial damage and halted Spitfire I production for a time. Fortunately, the Castle Bromwich factory was operational by then and turning out Spitfire IIs in increasing numbers. Eventually, Supermarine had no fewer than 46 small production and 19 support facilities in the south of England.

The Problem of Identification

Due to the myriad of Spitfire marks and models produced (along with their various engines, canopies, tail surfaces, armament fits and so on), there has always been some confusion about what some of the designation prefixes and suffixes mean. Indeed, it is entirely possible to have two similarly designated Spitfires looking very different and two apparently identical examples differently designated. In order to help clear up the confusion, the following explanation is offered.

The final letter in a Spitfire's designation (eg Spitfire F.VC) denoted the

type of wing the aircraft was fitted with, a function of its installed armament. The basic Spitfire elliptical wing had four derivatives – A, B, C, and E, while the Mark 21 and later models had an entirely new wing with a standard armament and no designation suffix. In the case of a model which was always fitted with the same wing, again no suffix was applied. An example is the Spitfire VIII.

'A' Wing: eight 0.303in Browning machine guns with 300 rounds per gun (rpg) on the Mk.IA; 350rpg on subsequent models.

'B' Wing: two 20mm Hispano cannon with 60rpg (Spitfire IB) and 120rpg (later models) plus four 0.303in Browning machine guns with 350rpg.

'C' Wing: or 'Universal' wing which was capable of accepting either the A or B armaments or four 20mm Hispano cannon. The vast majority had the B arrangement, identifiable by the blanked off cannon stubs outboard of the active guns.

'E' Wing: two 20mm Hispano cannon with 120rpg plus two 0.5in Browning machine guns with 250rpg; identifiable by the blanked off cannon stubs inboard of the active guns.

The Mark 21 and later models' new wing housed four 20mm cannon with 175rpg (inboard) and 150rpg (outboard).

The Royal Air Force introduced role prefixes in 1942. Before then, aircraft were designated simply Spitfire Mk.IA, for example. The prefixes relating to the Spitfire were as follows:

LF: optimised for low level operations with a low blown engine (supercharger gears set for maximum power at lower altitudes) and often with clipped wings, although it should be noted that the span of the wing fitted had no function in the designation.

F: designation for standard aircraft with engine supercharging set for medium/high altitudes; usually fitted with standard wingspan, although once again this had no influence on the designation.

HF: for high altitude operations with a high blown engine and sometimes with extended span wings.

FR: fighter/reconnaissance, usually with cameras and guns.

PR: photo/reconnaissance, usually unarmed.

The Merlin Spitfires

Following more than two years of testing and development with the prototype, the Spitfire finally went into production in 1938, the first Mk.IA being handed over to the RAF in July of that year. As it was first produced, the Spitfire I differed from the prototype in its triple

A photo call for the first RAF squadron equipped with the Spitfire – No 19 at Duxford. This photograph was taken in May 1939, nine months after the first aircraft was received.

Merlins and Fives

It's worth looking a little closer at these marks of the Merlin, because as they were applied to the Spitfire V which was delivered from early 1941, they represented the first time the different versions of the engine were tailored for specific roles.

The basis of these new marks of Merlin was the centrifugal supercharger which although still single-speed/single-stage units, allowed more power to be developed. The Merlin 45 was rated for medium altitudes (1,470hp/1,096kW) at 9,250ft), the Merlin 46 for higher altitudes (1,415hp/1,055kW) at 14,000ft thanks to an increased diameter supercharger impeller and the Merlin 45M was a specialised low altitude version with a cropped supercharger rotor. It was rated at 1,585hp (1,182kW) at only 2,750ft.

The Merlin 50 was similar to the 45 but featured the much sought after negative 'g' carburettor, the 50M bore a similar relationship to the 45M and the Merlin 55/56 was rated similarly to the 50 and 45 but had a two piece cylinder block and automatic boost control. The 55M was the low altitude version of the 55.

The Spitfire V was the most numerous mark to be built, some 6,487 of them coming off three production lines between the first in early 1941 and the last more than two years later. Combining the basic Mk.I airframe with the more powerful Merlins, three bladed de Havilland or Rotol propellers, strengthened fuselage longerons and more wing armour, the Spitfire V was built in three major versions. The first 94 aircraft were designated F.VA with

ejector exhausts and tailwheel. Early aircraft were powered by a Merlin II rated at 1,030hp (768kW) at 16,250 feet (4,955m) driving a two bladed fixed pitch Airscrew Company propeller. A similarly rated Merlin III soon became the powerplant of the Mk.IA, and in the quest for extra performance, three bladed propellers quickly became standard equipment, firstly a de Havilland two speed then a Rotol constant-speed unit. This gave the 'Battle of Britain' Spitfire a top speed of around 355mph (571km/h) at 19,000 feet (5,790m) and a climb performance of 9.4 minutes to 20,000 feet (6,095m).

Early production aircraft were identifiable (apart from the two blade propeller) by a flat cockpit canopy (soon replaced by a domed unit for more headroom) and a total of 1,567 was built including 30 experimentally fitted with cannon armament.

The Spitfire Mk.II was the first version to be built at the Castle Bromwich factory, delivery beginning in August 1940. The standard model was the Mk.IIA with improved armour protection for the pilot and a Merlin XII engine giving 1,175hp (876kW) for takeoff. Of the 921 Mk.IIs built, 170 were designated Mk.IIB with the standard 'B' wing mixed armament of two cannon and four machine guns and were therefore the first cannon armed Spitfires to enter regular service.

The Spitfire Mk.III was intended as the first Spitfire with a substantial

increase in performance. Powered by a Merlin XX with two-speed supercharger (giving 1,260hp/940kW at 11,750 feet) and featuring increased armour, reduced wing span and a retractable tail wheel, the Mk.III was ordered in quantity but only two prototypes were produced in 1940-41.

Following the conversion of many Spitfire Is to photo-reconnaissance configuration, the Spitfire PR.IV was purpose-built for that role was capable of flying 2,000 miles (3,218km) thanks to the removal of the armament and the fitting instead of long range fuel tanks in the wing leading edges; 229 were built. The PR.IV's airframe was based on that of the Spitfire V (see below) and power was provided by one of 40 or 50 series Merlins which also powered the Mk.V.

Spitfire fuselages under construction at Supermarine's Itchen plant in 1939. Walrus amphibians can be seen in the background.

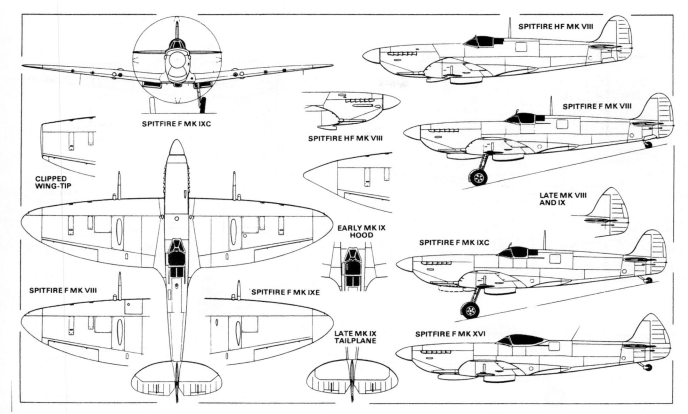

SPITFIRE HF MK VIII

SPITFIRE F MK IXC

SPITFIRE HF MK VIII

CLIPPED WING-TIP

SPITFIRE F MK VIII

EARLY MK IX HOOD

SPITFIRE F MK IXE

LATE MK IX TAILPLANE

SPITFIRE F MK VIII

LATE MK VIII AND IX

SPITFIRE F MK IXC

SPITFIRE F MK XVI

machine gun armament, followed by 3,911 F.VBs. The remainder were F.VCs with the universal wing and a further increase in armour protection. The vast majority of the latter version had the mixed cannon/machine gun armament although some were fitted with four cannon only.

All the Merlin models mentioned above were fitted to the Spitfire V at some stage, later aircraft having 50 series engines with negative 'g' carburettors, while another major operational improvement made during the course of the Mk.V's production run was the fitting of metal instead of fabric covered ailerons.

The Spitfire V's relatively long production life and high numbers built saw its specification achieve a number of firsts apart from those listed above. It was the first model to have its internal fuel capacity increased – to 113imp gal (514 l) on some late production aircraft by the fitting of a rear fuselage tank – the first to regularly operate with underfuselage drop tanks of up to 170imp gal (773 l) capacity; the first to be fitted with underwing bombs (usually as a field modification); and the first Spitfire to be tropicalised by the fitting of the bulbous Vokes filter under the nose, a necessary appendage for successful desert flying.

From 1943 most surviving Spitfire Vs were converted to LF.VA, LF.VB and LF.VC configuration with low altitude Merlin 45M, 50M or 55M engines and clipped wings (by simply removing the detachable wingtips and then covering the gap with a blank) for greater manoeuvrability at the low altitudes the aircraft were by now being flown in their daily sweeps across the English Channel and into France. 'Clipped, cropped and clapped' was the popular description given to these Spitfires.

In its standard form, a Spitfire VB was capable of a top speed of about 370mph (595km/h) at optimum altitude and despite increased weights was capable of climbing to 20,000 feet (6,096m) in 7.5 minutes, two minutes less than the original version. Tropicalisation reduced the Spitfire's performance slightly, although old age and the effects of constant repair had a rather more marked effect, some such aircraft recording top speeds 20mph (32km/h) slower than a factory fresh example.

While the Spitfire LF.V had been developed especially for low altitude roles, the Mk.VI (no role prefix or armament suffix was applied) was developed for the other end of the scale. Intended mainly to counter the high flying Junkers Ju 86P reconnaissance aircraft which appeared over Britain from 1941, the Spitfire VI was based on the airframe of the

Mk.V but fitted with a high rated Merlin 47 (1,415hp/1,055kW at 14,000ft), a lightly pressurised cabin (2.0psi) with sealed canopy, a four bladed Rotol propeller (the first Spitfire production version with this feature) and extended span wings housing the normal 'B' armament. One hundred were delivered to the RAF between April and October 1942.

The Two Stage Spitfires

For a 'temporary expedient', the Spitfire IX achieved a remarkable production run of 5,656 examples between 1942 and 1945 in numerous sub variants and configurations. Based on the Spitfire V airframe, the Mk.IX was intended as a mere stop-gap to counter the successful Focke-Wulf 190 fighter which had proved itself superior to the Spitfire V in the skies over Europe.

The definitive Merlin powered Spitfire was intended to be the Mk.VIII (see below) with a largely structurally redesigned airframe and the new two-speed, two-stage '60 series' engine, but the arrival of the Fw 190 in 1941 required a similarly powered but more quickly available Spitfire to be rushed into service as soon possible.

As with the Spitfire V, some explanation of the Rolls-Royce Merlin variants employed in the Mk.IX

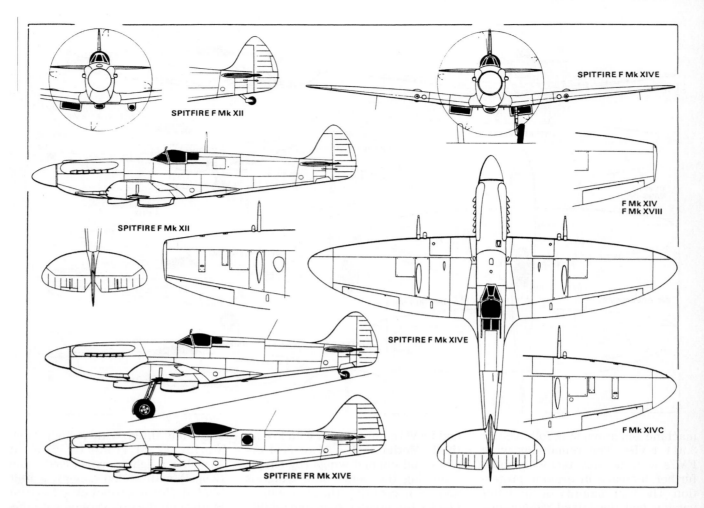

SPITFIRE F Mk XII

SPITFIRE F Mk XIVE

SPITFIRE F Mk XII

F Mk XIV
F Mk XVIII

SPITFIRE F Mk XIVE

SPITFIRE FR Mk XIVE

F Mk XIVC

might be necessary in an attempt to clear up what is an even more complicated scenario.

The basic difference between the Merlin 60 series engines and its predecessors was the incorporation of a two-speed/two-stage supercharger which meant that the fuel/air mixture was compressed by successive impellers, one in effect supercharging the other. The result was a substantial increase in power and rated altitude if required, plus even more options when it came to 'tailoring' the supercharger gear ratios to different role requirements.

Physically, the 60 series Merlins were longer engines than before and required a slightly lengthened engine cowling on the Spitfire plus a four bladed propeller to absorb the extra power. The new installation also featured multiple exhaust stubs in place of the previous triple ejector units, while the Spitfire IX lost the previous asymmetric look when the small oil radiator under the port wing was replaced with a combined oil cooler/supercharger intercooler which matched the engine cooling radiator under the other wing. The

60 series Merlins as fitted to the Spitfire featured two piece cylinder blocks, automatic mixture control and on later aircraft, interconnected propeller and throttle controls.

The engines fitted to the Spitfire IX were the medium/high rated Merlin 61 (1,565hp/1,167kW at 12,250ft and +15lb boost); the all-rounder Merlin 63/63A (1,710hp/1,275kW at 8,500ft and 1,510hp/1,126kW at 21,000ft with +18lb boost); the low/medium altitude Merlin 66 (1,705hp/1,271kW at 5,750ft and 1,580hp/1,178kW at 16,000ft) and the high altitude Merlin 70 (1,655hp/1,234kW at 10,000ft and 1,475hp/1,100kW at 22,500ft) with a Bendix injection carburettor installation developed by the Merlin's American licencee, Packard.

As its airframe was based on that of the Spitfire V, the Mk.IX featured a fixed tailwheel and all but the very last models had the standard Spitfire canopy and fin/rudder design. Later on, a cut down rear fuselage and 'bubble' canopy were introduced, along with a more pointed fin and rudder.

The initial versions produced were the F.IXC (Merlin 61 or 63 and stand-

ard wings); LF.IXC (Merlin 66, some with clipped wings but many converted to that configuration later); and HF.IXC (Merlin 70, most with the standard wing). In early 1944 the 'E' wing with its two cannon and four 50-calibre machine guns was introduced and F.IXE, LF.IXE and a handful of HF.IXE models were produced.

The PR.IX was an unarmed conversion of the fighter (15 converted), while the FR.IX was also a conversion performed in small numbers in which the armament was retained. Supermarine phased out production of the Spitfire IX in mid 1943 but Castle Bromwich continued building them until two years later. The final split of variants built was 4,010 LF.IXs, 1,246 F.IXs and 400 HF.IXs.

Installation of the new series of Merlins transformed the Spitfire's performance, and for the first time it was able to exceed 400mph (644km/h) in level flight. Its initial rate of climb of better than 4,000 feet per minute (1,220m/min) and a service ceiling of around 43,000ft (13,100m) maintained the Spitfire's reputation as a top class interceptor,

The wing's the thing – three Spitfires demonstrate variations on the characteristic elliptical wing. Spitfire VB R6923 (converted from a Mk.I) of 92 Squadron (above left) with standard wing; LF.VB with clipped wing (above right); and Mk.VI prototype X4942 (converted from a Mk.V) with extended tips for high altitude work (below).

although its range on internal fuel was still poor despite a modest increase in internal capacity.

Following the Mk.IX into service were the Mks.VII and VIII. These versions represented the first substantial redesign of the Spitfire and both had the new two-speed/two-stage Merlins, four bladed propeller, symmetrical radiators, longer nose and other external features of the Mk.IX but in combination with a considerably structurally redesigned and stronger fuselage and revised 'C' wing with shorter span ailerons. A retractable tail wheel was also fitted and some fuel in the wing leading edge increased the internal capacity and the range to over 600 miles (965km).

The Spitfire VII was a specialised high altitude variant powered by either a Merlin 61/64 (F.VII) or Merlin 71 (HF.VII), the latter engines similar to the Merlin 63 and 70 but with the facility for a cabin blower to feed the Spitfire VII's lightly pressurised cabin. All Mk.VIIs featured the extended span wing (and could comfortably achieve 44,000ft) and most

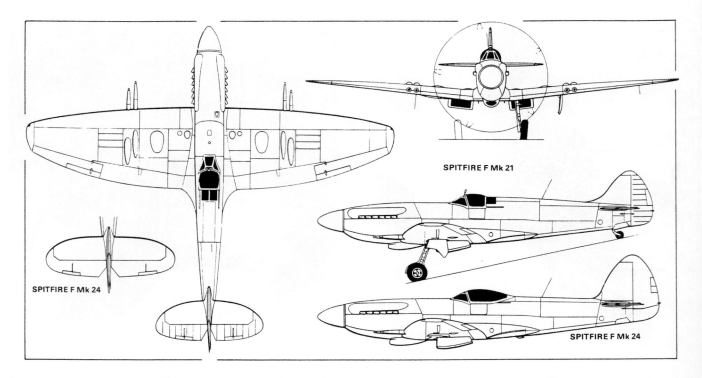

SPITFIRE F Mk 21

SPITFIRE F Mk 24

SPITFIRE F Mk 24

of the 140 delivered from September 1942 had the standard fin and rudder shape.

The Spitfire VIII has been generally reckoned to represent the pinnacle of Merlin-Spitfire design. Basically an unpressurised version of the Mk.VII and incorporating all of its new features, 1,652 were built in F.VIII, HF.VIII and LF.VIII variants between late 1942 and the end of 1944.

Most were built with the pointed fin and rudder and standard span wings (despite their role prefixes, the variation applying only to the type of Merlin fitted), while all were tropicalised on the production line, with a slightly larger than standard Aero-Vee carburettor intake under the nose.

Various Spitfire VIIIs were used for a number of experiments including the prototype installation of the cut down rear fuselage and bubble canopy, counter rotating propellers and a laminar flow wing. The production total excludes six examples which were completed as prototypes for the Griffon powered Spitfire XIV.

The Spitfire PR.X and PR.XI were basically unarmed photo-reconnaissance versions combining features of the Mks.VII, VIII and IX, but with additional fuel capacity. Only 16 pressurised PR.Xs were built in 1944, but PR.XI production reached 471 examples from late 1942.

The final major Merlin powered Spitfire variant was the Mk.XVI which was built during the last year of the war and for a few months afterwards. This was simply a Spitfire IX powered by a low/medium altitude Packard-Merlin 266 (equivalent to the Rolls-Royce built Merlin 66) but later models featured an extra fuselage ferry tank.

The Spitfire XVI was built with both the 'C' (most with clipped tips, standard canopy and fin) and 'E' wings (mostly with clipped tips, bubble canopy and all with pointed fins). A total of 1,054 of the final Merlin engined Spitfire variant was built and all were designated LF.XVIs.

The Griffon Spitfires

The never ending search for extra power and therefore speed in fighter aircraft of the 1930s and 1940s resulted in early consideration being given to re-engining the Spitfire with the Rolls-Royce Griffon engine, a powerplant which shared the liquid cooled vee-12 layout of the Merlin but with a cubic capacity some one-third greater. The use of the Griffon was being considered as early as 1939, but it would not be until November 1941 that a Spitfire would fly with this potentially much more powerful engine installed.

A development of the Rolls-Royce R racing engine which had powered the Schneider Trophy winning Supermarine floatplanes, the Griffon had been a low key project at Rolls-Royce during the 1930s as the

For a 'temporary expedient', the Spitfire IX with Merlin 60 series engine with two-speed/ two-stage supercharger was built in very large numbers. These early Mk.IXs are from No 611 'West Lancashire' Squadron, one of the first units to receive this important Spitfire variant.

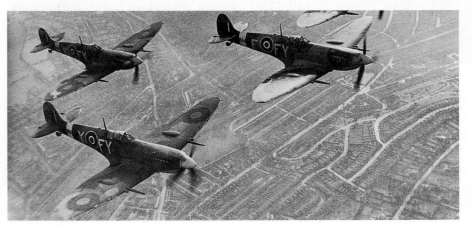

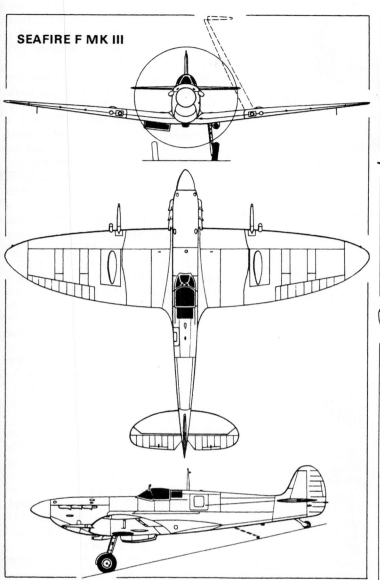

SEAFIRE F MK III

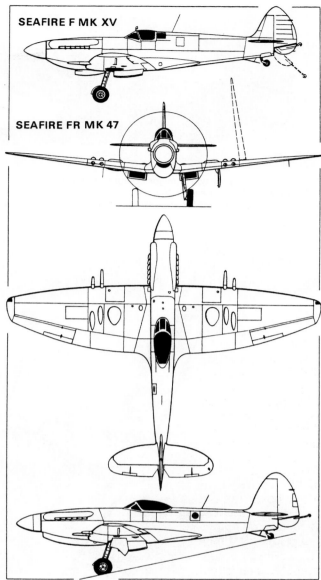

SEAFIRE F MK XV

SEAFIRE FR MK 47

Merlin had taken precedence. But with official interest being shown as 1940 dawned, development was restarted and a Griffon powered Spitfire would soon be flying.

Apart from its larger capacity, the Griffon differed from the Merlin in several ways. Its rotational direction was opposite to that of the smaller engine, it was longer, and it had a lower thrust line which meant that a completely new engine cowling shape (with the cylinder blocks outline obvious) would grace Spitfires thus powered.

Originally designated Spitfire IV and later Spitfire XX, the first Griffon-Spitfire was built in prototype form only, using a modified Mk.I airframe as its basis. Externally it differed by virtue of its new nose shape, retractable tail wheel and four bladed propeller. The installed two-speed/

single-stage Griffon IIB was rated at 1,735hp (1,294kW) and following its first flight in November 1941 the aircraft immediately exhibited a great improvement in performance, reaching a top speed of 433mph (697km/h) at 23,000ft (7,010m).

The Spitfire XII was the first Griffon powered model to reach production after flying for the first time in August 1942. The first of 100 reached the RAF in February 1943. Power was provided by a 1,735hp (1,294kW) Griffon III or IV optimised for low level operations, the wing was a clipped 'C' unit, the standard canopy was fitted as was the pointed fin which was beginning to appear on other Spitfires. The XII's top speed was 393mph (632km/h) at 18,000ft (5,484m) and it was one of the fastest in the world down low at the time.

With the standard Spitfire I/II/V fuel capacity of 85imp gal (386 l) its range was poor, although this was not a major problem given its intended role – as a point interceptor to counter Fw 190 'hit and run' raids on England's south coast. Early aircraft were based on the Spitfire V airframe with its fixed tail wheel; later examples had the Mk.VIII airframe as their basis and therefore a retractable tailwheel.

The Griffon-Spitfire concept was refined with the next version, the Mk.XIV. Although intended as an interim model, true to form, it became the most produced of all the Griffon powered Spitfires.

This time the powerplant was the useful at all altitudes two-speed/two-stage Griffon 65 rated at 2,050hp (1,528kW) at 7,000ft (2,130m) and 1,820hp at 21,000ft

Photo-reconnaissance was a major operational role for the Spitfire, many specialist versions being produced from new or by conversion. This is an example of the most numerous, the PR.XI. Note the enlarged lower cowling to house a bigger engine oil tank, necessary for long endurance flights.

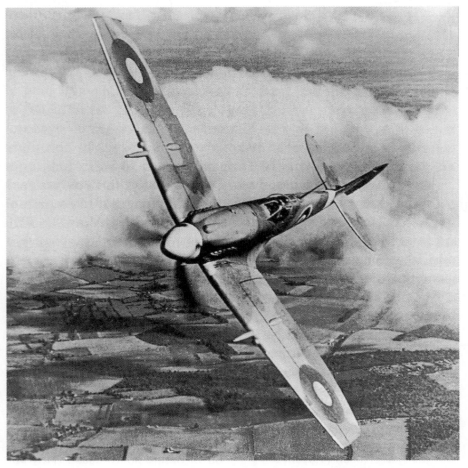

(above) The Mk.XII was the first production version of the Griffon powered Spitfire, entering service in early 1943. (below) Installing the larger capacity Rolls-Royce Griffon engine in the Spitfire bestowed the aircraft with very high performance. DP845 was the first Griffon powered prototype Mk.IV (later Mk.XX), first flying in November 1941.

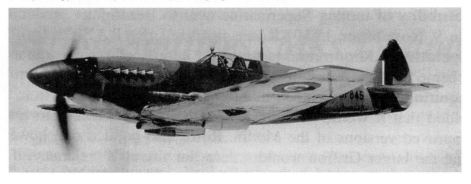

(6,400m); a five bladed Rotol propeller was fitted, as was a new broader chord fin and rudder and the smaller span ailerons and retractable tail wheel of the Mk.VIII.

Within the 957 production Spitfire XIVs delivered between January 1944 and the end of 1945 there were three subvariants: the F.XIV with a standard 'C' wing and standard canopy; F.XIVE with standard or clipped 'E' wing and bubble canopy in the later models; and FR.XIVE fighter/reconnaissance version with clipped wings, bubble canopy, a camera and a rear fuselage fuel tank.

As the last Spitfire variant to enter full service in World War II, the Mk.XIV also took the type's performance to a level which was at least a match for any other piston engined fighter of the time. Top speed was 448mph (721km/h) at 25,400ft, 20,000ft could be reached just five minutes after takeoff, maximum rate of climb was a prodigious 4,580ft (1,396m) per minute and service ceiling was over 44,000ft (13,400m).

Further development of the Spitfire XIV resulted in the F and FR.XVIII which was similar to a late model Mk.XIVE but with a strengthened wing and extra rear fuselage fuel tanks (300 built from March 1945) and the PR.XIX which shared the external features and mechanical specification of the Mk.XIV combined with the 'wet' wing and camera installation of the PR.XI.

Of the 225 PR.XIXs delivered from May 1944, the first 22 were powered by a Griffon 65 and unpressurised, the remaining 200 had a pressurised cabin, a similarly rated Griffon 66 and a further increase in fuel capacity.

The final three marks of Spitfire bore little relationship to the earlier models as they incorporated an entirely new wing fitted with four 20mm cannon. The first was the Spitfire F.21 (Arabic appellation replaced Roman in the RAF after 1945) with the new wing, a 2,050hp (1529kW) Griffon 61 or 64, a larger diameter Rotol five bladed propeller and lengthened undercarriage struts.

The first true prototype first flew in July 1943 but the F.21 didn't enter service until early 1945 after a lengthy series of tests and trials became necessary to cure severe stability problems. Production models were identifiable by their square wing tips, broad rudder, standard canopy and fully enclosed wheel

wells; 120 production F.21s were built.

The final Spitfires built were the Mks.22 and 24. Based on the F.21 and similarly powered, they differed by having bubble canopies and the very large fin and rudder intended for the Spitfire's unsuccessful follow up, the Spiteful. The first Spitfire F.22 flew in March 1945, the final F.24 (which differed in having more fuel and electrical instead of pneumatic gun firing) was rolled off Supermarine's production line in February 1948. Between them, the Spitfire 22 and 24 added 342 to the production tally.

Sea Spitfires

Investigations into a carrier borne 'Sea Spitfire' began in 1939 but the RAF's soon to be urgent needs meant the first 'Hooked Spitfire' (a converted Mk.V) did not fly until 6 January 1942.

In service, what was called the Seafire provided the Royal Navy Fleet Air Arm with an excellent fighter that was able to match most adversaries it was likely to encounter, although the aircraft's narrow track undercarriage caused some landing and deck handling difficulties. Seafires flew with 25 FAA operational squadrons from 1942 to 1950, its combat debut recorded during the Allied invasion of North Africa (Operation Torch) in late 1942. It subsequently operated in Sicily, Norway, the D-Day landings, Europe and the Far East, plus post war active service in Korea.

The initial 166 Seafire IBs were converted from tropicalised Spitfire VBs by Air Service Training and delivered from June 1942. These aircraft featured the standard Merlin 45 or 47 powerplant, vee frame arrester hook and catapult spools. They were allocated new serial numbers when converted.

The first new build model was the Seafire II, based on the Spitfire VC (retaining non folding wings) but with catapult attachment points, arrester hook, strengthened undercarriage and provision for rocket assisted takeoff gear (RATOG). Production reached 372 by Supermarine and Westland in three versions: F.IIC with 1,470hp (1,096kW) Merlin 45 or 1,415hp (1,055kW) Merlin 46 and three bladed propeller; L.IIC with low rated 1,645hp (1,226kW) Merlin 32 or 1,585hp (1,182kW) Merlin 55M; and the similarly powered LR.IIC with cameras.

Most Seafire IIs had clipped wings, some the Vokes tropical filter and deliveries began in June 1942, the aircraft largely converted from Spitfire VCs on the production line.

The fully navalised Seafire III with folding wings and four bladed propeller was introduced to service in April 1943 and manufactured until May 1945, Westland and Cunliffe-Owen Aircraft between them building 1,220 aircraft. Variants were the F.III (Merlin 55); LF.III (low rated Merlin 32 or 55M, most with clipped wings); and FR.III with cameras. An additional 30 were built without folding wings under the designation Mk.III (Hybrid).

Development of Griffon powered Seafires was also pursued, beginning with the Seafire F.XV, based on the Spitfire XII with folding 'C' wing and the naval features of the Mk.III. Powered by an 1,815hp (1,353kW) Griffon VI driving a four bladed propeller, the first prototype flew in December 1943 but service entry wasn't until May 1945, too late to see wartime service. Production reached 384 by Westland and Cunliffe-Owen plus six prototypes by the parent company.

The Seafire XVIII was an improved version of the F.XV, still powered by a Griffon VI but with a 'bubble' canopy on later aircraft, strengthened and longer stroke undercarriage, rear fuselage fuel tank and 'sting' type tail hook as also used on subsequent models. The Seafire XVII entered service in September 1945, production reaching 232 by Westland and Cunliffe-Owen as the F.XVII although some were subsequently converted to FR.XVIIs with a camera in place of the rear fuel tank.

The final Seafire Mks.45, 46 and 47 were based on the rewinged Spitfire 21, 22 and 24 respectively, powered by two-speed/two-stage Griffons and armed with four 20mm cannons. All three were built in F (fighter) and FR (fighter/reconnaissance) versions, the latter sometimes by conversion.

The Mk.45 was powered by a 2,035hp (1,517kW) Griffon 61 driving a five bladed propeller or 2,045hp (1,525kW) Griffon 85 with six bladed Rotol contra-props on some aircraft. The wings were non folding, the standard canopy was fitted and the undercarriage was fully enclosed by doors when retracted. Fifty were built at Castle Bromwich and delivered between March 1945 and January 1946 as the only Seafires manufactured at that facility and the last of the Spitfire/Seafire line to emerge from it.

The Seafire 46 also had non folding wings plus the bubble canopy and the enlarged vertical tail surfaces developed for the Spiteful, these features along with the new wing at last removing any real physical resemblance to the classic Spitfire lines. Five bladed propellers or six bladed contra-props were fitted driven by various Griffons of up to 2,375hp (1,771kW). Additional fuel capacity was provided and only 24 were built.

The Seafire 47 represented the ultimate development of the Spitfire/Seafire line with 90 built between April 1946 and the final delivery on 28 January 1949. Powered by a 2,145hp (1,599kW) Griffon 87 or 2,375hp (1,771kW) Griffon 88, all Seafire 47s had six bladed contra-props and folding wings. The last front line RN FAA squadron to fly Seafires was No 800, which embarked its Mk.47s on HMS *Triumph* in 1950.

SPITFIRE and SEAFIRE PRODUCTION

SPITFIRE

Mark	Supermarine	CBAF	Westland	Total
Prototype	1	–	–	1
I	1517	–	50	1567
I (PR)	2	–	–	2
II	–	921	–	921
III	2	–	–	2
IV/XX	2	–	–	2
PR.IV	229	–	–	229
V	1363	4489	635	6487
VI	100	–	–	100
VII	140	–	–	140
VIII	1652	–	–	1652
IX	561	5095	–	5656
X	16	–	–	16
XI	471	–	–	471
XII	100	–	–	100
XIV	963	–	–	963
XVI	–	1054	–	1054
XVIII	300	–	–	300
XIX	225	–	–	225
21	1	120	–	121
22	28	260	–	288
24	54	–	–	54
Totals	**7,727**	**11,939**	**685**	**20,351**

Spitfire I breakdown: 1537 Mk.IA, 30 Mk.IB.
Spitfire II breakdown: 751 Mk.IIA, 170 Mk.IIB.
Spitfire V breakdown: 94 Mk.VA, 3911 Mk.VB, 2467 Mk.VC, 15 PR Type F.
Spitfire VIII breakdown: 267 F.VIII, 160 HF.VIII, 1225 LF.VIII.
Spitfire IX breakdown: 4010 LF.IX, 1246 F.IX, 400 HF.IX.
Spitfire XIV breakdown: 527 F.XIV, 430 FR.XIV.
Spitfire XVIII breakdown: 100 F.XVIII, 200 FR.XVIII.

SEAFIRE

Mark	Supermarine	CBAF	Westland	Cunliffe-Owen	Total
II	262	–	110	–	372
III (Hybrid)	–	–	30	–	30
III	–	–	870	350	1220
XV	6	–	250	134	390
XVII	–	–	212	20	232
45	–	50	–	–	50
46	24	–	–	–	24
47	90	–	–	–	90
Totals	**382**	**50**	**1,472**	**504**	**2,408**

Seafire III breakdown: 100 F.III, 129 FR.III, 991 L.III.

(above) An early production Spitfire F.22 with various panels still to be installed before completion at Castle Bromwich. This aircraft has the original relatively small horizontal and vertical tail surfaces – later aircraft had enlarged surfaces. (below) Navalised Spitfires were called the Seafire and built in both Merlin and Griffon powered versions. This is a Merlin powered Seafire III, the first model to feature folding wings.

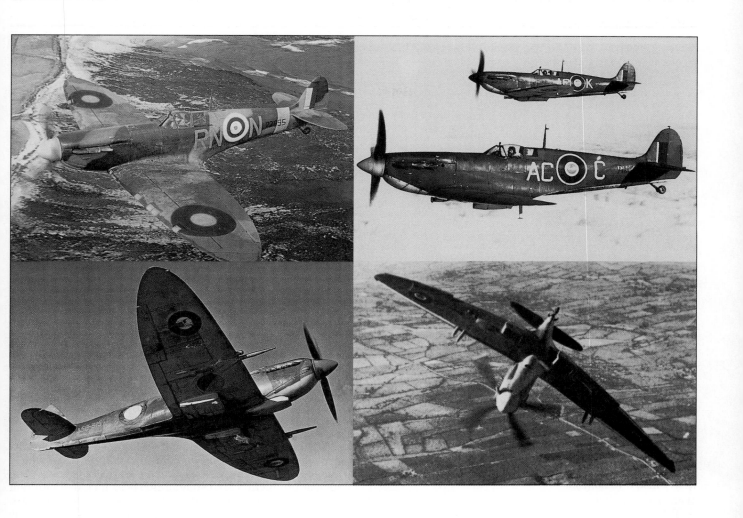

MARKS and MODELS

The prototype Spitfire photographed in May 1936 (left) and later in its career (below) when brought up to Mk.I standards.

Type Number: 300.

First flight: 5 March 1936 (K5054).

Service entry: Not applicable.

Powerplant: 990hp (738kW) Rolls-Royce Merlin C with 1-speed/1-stage supercharger and +6lb max boost; 1,035hp (772kW) Merlin F; or 1,030hp (768kW) Merlin II; two bladed fixed pitch propeller initially then Fairey-Reed three bladed fixed pitch, de Havilland three bladed two-pitch. Internal fuel 75imp gal (341 l) in forward fuselage tanks.

Dimensions: Wing span 36ft 10in (11.23m); length 29ft 11in (9.12m); height (propeller vertical) 12ft 8in (3.86m); wing area 242sq ft (22.5m²).

Weights: May 1936 – empty 4,082lb (1,852kg); loaded 5,359lb (2,431kg).

Armament: Originally none, eight 0.303in Browning machine guns with 300rpg from December 1936.

Performance: Merlin C, two bladed propeller – max speed 349mph (562km/h) at 16,800ft (5,120m), 330mph (531km/h) at 10,000ft (3,050m); max climb 2,920ft (890m)/min; time to 20,000ft (6,095m) 8.2min; time to 30,000ft (9,145m) 17.0min; service ceiling 35,400ft (10,790ft).

Production: 1.

Notes: The Spitfire's origins date back to October 1931 and the issuing by the Air Ministry of Specification F.7/30, which resulted in the unsuccessful Supermarine Type 224 fighter, the sole example of which was first flown in February 1934. A low winged design with fixed undercarriage powered by an evaporative cooled Rolls-Royce Goshawk engine, the 224 was armed with four 0.303in machine guns but its top speed was only 238mph (383km/h). This and its rate of climb was inferior to the competing and more manoeuvrable Gloster SS 37 biplane which was awarded a production contract.

Supermarine's Reginald Mitchell began working on the Type 300 monoplane fighter powered by Rolls-Royce's new P.V.12 (Merlin) engine shortly after the 224 was flown. The design evolved through several phases after the original proposals were made in July 1934. The famous elliptical wing planform first entered the equation in November 1934 and in December 1934 production of a single prototype was ordered. It was originally intended to have a four gun armament but Mitchell was able to prove his wing design could carry eight guns.

The name 'Spitfire' was first used in mid 1935 and the prototype (K5054) first flew on 5 March 1936 with J 'Mutt' Summers at the controls.

Initial testing of the prototype revealed performance to be disappointing, the maximum speed of 338mph (544km/h) at optimum altitude well below the target of 350mph (563km/h). Propeller and exhaust system development quickly boosted this to the required figure and early service testing of the aircraft resulted in high praise of its handling qualities from pilots.

The prototype was continually developed throughout its 3½ years of its existence with aerodynamic and mechanical changes being incorporated which eventually brought it up to near production Mk.I standards. These included different versions of the Merlin, different propellers, modified flaps, replacement of the tail skid with a wheel, removal of the main undercarriage wheel covers and modified rudder. Armament was installed in December 1936, testing quickly revealing a problem with the guns freezing, a flaw which took about two years to sort out.

After surviving several relatively minor accidents and incidents, K5054 was finally written off on 4 September 1939 – the day after war was declared – when it turned over on landing. It had logged about 260 flying hours.

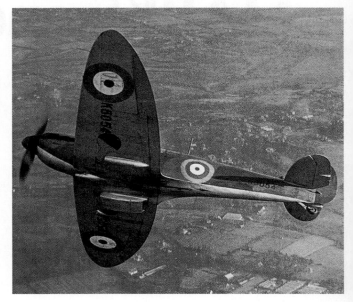

Spitfire I

Spitfire IAs of 65 Squadron (top) with the nearest aircraft flown by Battle of Britain ace Robert Stanford Tuck; early production Spitfire I K9845 (bottom left) with two bladed propeller; and Mk.IA X4913 (bottom right).

Type number: 300.

First flight: 14 May 1938 (K9787).

Service entry: August 1938.

Powerplant: 1,030hp (768kW) Rolls-Royce Merlin II or III with 1-speed/1-stage supercharger and +6¼lb max boost; two bladed fixed pitch, three bladed de Havilland two-speed or three bladed de Havilland or Rotol constant-speed propeller. Internal fuel 85imp gal (386 l) in forward fuselage tanks.

Dimensions: Wing span 36ft 10in (11.23m); length 29ft 11in (9.12m); height 11ft 5in (3.48m); wing area 242sq ft (22.5m²).

Weights: Typical IA – empty 4,810lb (2,182kg); loaded 5,844-6,200lb (2,651-2,812kg).

Armament: IA – eight 0.303in Browning machine guns in wings with 300rpg. IB – two 20mm Hispano cannon with 60rpg; some with additional four 0.303 Browning machine guns.

Performance: Late production Mk.IA – max speed 363mph (584km/h) at 18,500ft (5,640m), 295mph (475km/h) at 2,000ft (610m); economical cruise 210mph (338km/h) at 20,000ft (6,095m); long range cruise 160mph (257km/h) at 14,000ft (4,270m); max climb 2,530ft (771m)/min; time to 20,000ft (6,096m) 9.4min; service ceiling 31,900ft (9,723m); max range 575 miles (925km).

Production: 1,567 (Supermarine 1,517, Westland 50).

History: Supermarine received its first order for production Spitfires on 3 June 1936, this initial contract covering 310 aircraft. By the time Britain declared war on Germany on 3 September 1939, 2,160 were on order and 306 delivered. Production was running at the rate of one per day at the time but deliveries were behind schedule due to difficulties in manufacturing the relatively complex airframe.

The first production Spitfire I flew on 14 May 1938 and the first operational squadron to receive the aircraft was No 19 at Duxford on 4 August.

Early aircraft were powered by the Merlin II engine driving a two bladed fixed pitch propeller. The similarly rated Merlin III with a three bladed two-pitch de Havilland propeller quickly followed and then a three bladed constant-speed unit. The improved performance this conferred resulted in early aircraft being converted to the new propeller in 1940. By August 1940, 19 RAF Fighter Command squadrons were equipped with Spitfire Is. Its first 'kills' where recorded on 16 October 1939 when two Ju 88s were shot down over the Firth of Forth in Scotland.

Many detail improvements were introduced in 1938-40 including replacing the original 'flat' canopy with a domed unit for increased pilot headroom (from the eighth production aircraft), increased armour protection and fitting a gun camera.

The machine gun armed Spitfire I was retrospectively designated Mk.IA with the appearance of the Mk.IB, initially with two 20mm cannon and no machine guns and then with the definitive 'B' wing with mixed armament. Trials with the cannon began in March 1939 and some flew with this armament during the Battle of Britain but stoppages were a problem. The first aircraft with the 'B' wing flew on 20 August 1940. Mk.IB production reached 30 aircraft.

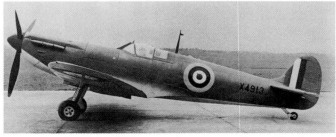

Spitfire IIA P7350 in 266 Squadron RAF markings.

Type number: 329.

First flight: 24 September 1939 (K9788 converted from Mk.I).

Service entry: August 1940.

Powerplant: 1,175hp (876kW) Merlin XII with 1-speed/1-stage supercharger and +12½lb max boost; de Havilland or Rotol three bladed propeller. Internal fuel 85imp gal (386 l) in forward fuselage tanks.

Dimensions: Wing span 36ft 10in (11.23m); length 29ft 11in (9.12m); height 11ft 5in (3.48m); wing area 242sq ft (22.5m²).

Weights: Empty 4,900lb (2,223kg); normal loaded 5,900lb (2,676kg); max loaded 6,317lb (2,865kg).

Armament: IIA – eight 0.303in Browning machine guns with 300rpg. IIB – two 20mm Hispano cannon (60rpg) and four 0.303in Browning machine guns (300rpg).

Performance: Max speed 357mph (575km/h) at 17,000ft (5,180m), 290mph (467km/h) at sea level; max cruise 310mph (499km/h) at 20,000ft (6,095m); economical cruise 195mph (314km/h); max climb 2,995ft (913m)/min; time to 20,000ft (6,095m) 7.0min; service ceiling 33,900ft (10,330m); max range 500 miles (805km).

Production: 751 Mk.IIA and 150 Mk.IIB by CBAF, total 921.

Spitfire IIC (ASR.II) showing loading of rescue packs. (Neil Mackenzie)

Notes: The Spitfire II was the first version built by the Castle Bromwich Aircraft Factory (CBAF) at Birmingham, the plant which contributed more than half of the total Spitfire/Seafire production run. The prototype (converted from a Mk.I) was first flown in September 1939 while the first production model to emerge from the new factory appeared in June 1940.

There were many teething problems at the CBAF which resulted in delays. Management was originally controlled by the Nuffield Organisation but the new Minister of Aircraft Production, Lord Beaverbrook, ordered changes immediately on his appointment with the result that Vickers-Armstrong took over management of the facility. No 611 Squadron at Digby was the first operational unit to receive the new aircraft, in August 1940, with the result that a small number saw action during the Battle of Britain.

Compared to the Spitfire I, the Mk.II featured the more powerful Merlin XII engine, its use of 100 rather than 87 octane fuel allowing higher supercharger boost pressures and therefore greater power outputs. Other differences included the use of a Coffman cartridge rather than electrical starting system for the engine and the incorporation from scratch of increased armour protection for the pilot.

One hundred and seventy Spitfire IIs were completed as Mk.IIBs with the two cannon/four machine gun 'B' wing. Compared to the earlier Spitfire IB with cannon, this version was the first to feature a revised feed and ejection system for the 20mm ammunition. In addition, the guns were mounted on their sides which introduced the characteristic 'B' wing blister on the upper wing: the result was a reliable and practical cannon installation which solved the jamming problems encountered by the Mk.IB.

Fifty Spitfire IIs were converted to Mk.IIC (later ASR.II) air-sea rescue aircraft in 1942-43 to meet a requirement for such an aircraft to operate in the Thames Estuary and English Channel areas. These had the capability to drop flares, rescue packs (including a dinghy and rations) or smoke bombs from chutes in the lower rear fuselage. Smoke marker bombs could also be carried on underwing racks.

In addition, these aircraft also swapped their standard Merlin XII for the more powerful 1,460hp (1,089kW) Merlin XX with 2-speed/1-stage supercharger. The ASR.II served with five search and rescue squadrons.

Spitfire III

Spitfire III prototype N3297 in its original clipped wing form (right and below left) and with standard wing tips late in 1940 (below right).

Type numbers: 330 and 348.

First flight: 15 March 1940 (N3297).

Service entry: Not applicable.

Powerplant: 1,460hp (1,089kW) Merlin XX with 2-speed/1-stage supercharger and +16lb max boost; three bladed Rotol propeller. Internal fuel 99imp gal (450 l) in forward fuselage tanks. Later testing – 1,565hp (1,167kW) Merlin 61 with 2-speed/2-stage supercharger and +15lb max boost; four bladed Rotol propeller.

Dimensions: As originally flown – wing span 30ft 6in (9.30m); length 30ft 4in (9.25m); height 11ft 5in (3.48m); wing area 220sq ft (20.4m^2).

Weights: 5,229lb (2,372kg); loaded up to 7,455lb (3,382kg).

Armament: Proposed – either 'A' (eight 0.303in Browning machine guns), 'B' (two 20mm Hispano cannon and four 0.303in machine guns) or 'C' (as A/B or four 20mm cannon).

Performance: Merlin XX – max speed 400mph (644km/h) at 21,000ft (6,400m), 340mph (547km/h) at 5,000ft (1,525m); time to 20,000ft (6,095m) 6.4min; service ceiling 38,000ft (11,580m).
Merlin 61 – max speed 414mph (666km/h) at 27,200ft (8,290m), 345mph (555km/h) at 4,000ft (1,220m), 391mph (629km/h) at 16,000ft (4,875m); max climb 2,900ft (884m)/min; service ceiling 41,800ft (12,740m).

Production: 2 by Supermarine.

Notes: The Spitfire III was the first extensively modified and upgraded development of the basic design. First proposed by Supermarine in February 1939, this 'Improved' Spitfire featured a number of substantial modifications.

The heart of the aircraft was the more powerful Merlin XX engine with two-speed supercharger, its installation requiring strengthening of the fuselage and modifications to the engine mounting and cowling. The wings were notably reduced in span, a retractable tailwheel was fitted, fuel capacity increased, armour protection enhanced, the oil cooler intake and coolant radiator under the wings increased in size, the strengthened main undercarriage was raked forward slightly and fairings covered the main wheels when they were retracted. The Mk.III was also the first Spitfire intended to have the universal 'C' wing fitted with its various armament options and also would have been the first production variant to reach 400mph (644km/h) in level flight.

Supply of the Merlin XX was initially a problem as this engine had been allocated to the Hurricane II, but with that supply guaranteed, an order for 1,000 Spitfire IIIs was placed with the Castle Bromwich Aircraft Factory in October 1940.

The prototype (N3297, converted on the production line from a Mk.I) first flew on 15 March 1940 but shortly after the production order had been placed the decision was taken to go ahead with the Merlin 45 powered Spitfire V instead of the more complex Mk.III at a time when minimal production disruption was important. The 1,000 on order were completed as Mk.Vs. N3297 was therefore used for tests and trials and made a major contribution to the Spitfire development programme when it emerged in September 1941 fitted with a Merlin 61 engine, four bladed propeller and standard span wings as in effect the prototype for the Mk.IX.

A second Spitfire III (W3237, converted from a Mk.V airframe) was flown in June 1941 and used for trials with extended wingtips, roll performance tests and for Seafire development conducting takeoff performance trials with various flap settings.

Spitfire IV/XX

Spitfire IV first prototype DP845 in its original form (left) with six cannon mockup and later (bottom left) with normal armament and broad chord rudder. The second Mk.IV (DP851, bottom right) is shown here after conversion to the prototype Mk.21 in late 1942.

Type number: 337.

First flight: 27 November 1941 (DP845).

Service entry: Not applicable.

Powerplant: 1,735hp (1,294kW) Rolls-Royce Griffon IIB with 2-speed/1-stage supercharger and +12lb max boost; four bladed Rotol propeller. Internal fuel 130imp gal (591 l) in forward fuselage and wing leading edge tanks.

Dimensions: Wing span 36ft 10in (11.23m); length 30ft 6in (9.30m); wing area 242sq ft (22.4m²).

Weights: With proposed six 20mm cannon – empty 6,439lb (2,921kg); loaded 9,150lb (4,150kg). With 'C' wing – empty 5,600lb (2,540kg); loaded 7,310lb (3,316kg).

Armament: Original proposal – either six 20mm cannon, four 20mm cannon or 12 0.303in machine guns. Prototype fitted with six 20mm cannon mockup and subsequently 'C' layout of two 20mm Hispano cannon and four 0.303in Browning machine guns.

Performance: Max speed 409mph (658km/h) at 18,600ft (5,670m), 354mph (570km/h) at sea level, 397mph (639km/h) at 25,000ft (7,620m); time to 15,000ft (4,570m) 4.5min; time to 20,000ft (6,095m) 6.6min; service ceiling 38,000ft (11,580m).

Production: 2 by Supermarine.

Notes: Rolls-Royce began development of what would become the Griffon engine in January 1939, the powerplant sharing the vee-12 configuration of the Merlin but with 36 per cent greater capacity (36.7 versus 27 litres). Despite its larger displacement, Rolls-Royce's engineers were able to keep the Griffon's frontal area to a figure only 6 per cent greater than the Merlin's and the length within 3in (7.6cm).

Applying the new engine to the Spitfire was an obvious course of action, first investigations into this occurring in November 1939. With official encouragement, Supermarine issued a specification for a Griffon powered Spitfire the following month, the aircraft incorporating a similar airframe to the Mk.III with features such as a retractable tailwheel, strengthened and raked forward main undercarriage legs, general structural strengthening and increased internal fuel capacity. A new spilt flap design was specified and planned armament options were six 20mm cannon, four 20mm cannon or 12 0.303in machine guns.

Work continued during 1940 but by the end of the year there was still no formal contract in place. An order for two prototype Spitfire IVs (as the new version was dubbed) was finally placed in May 1941 and a production contract for 750 placed in August of the same year. As it happened, no production Mk.IVs were built, and the contract was amended to cover a similar number of Spitfire Vs which were regarded as being more urgently needed.

The prototype Spitfire IV (DP845) first flew on 27 November 1941, fitted with standard span wings and a six cannon mockup installation. The aircraft was redesignated as the Mk.XX from March 1942 to avoid confusion with the PR.IV.

Testing revealed DP845 to be one of the fastest fighters in the world at the time, easily outpacing a Hawker Typhoon and a captured Focke-Wulf 190 during testing. With Griffon-Spitfire production plans temporarily changed, DP845 went on to a career as a testbed with clipped wings and modified fin and rudder and also flew as the prototype for the Spitfire XII in August 1942. A second Mk.IV/XX (DP851) was flown in December 1941, this serving as a development aircraft for the Spitfire 21.

Spitfire Mk.I(PR), PR.IV, PR.VII and PR.XIII

Type numbers: PR.IV – 353. PR.XIII – 367.

First flights: Original Mk.I conversion (N3069) – October 1939. Type D/PR.IV (P9551) – September 1940. PR.XIII (L1004) – August 1942.

Service entry: See notes.

Powerplants: Mk.I conversions – 1,030hp (768kW) Rolls-Royce Merlin III. PR.IV/VII – 1,470hp (1,096kW) Merlin 45/50/55/56 or 1,415hp (1,055kW) Merlin 46, all with 1-speed/1-stage supercharger and +16lb max boost. PR.XIII – 1,645hp (1,227kW) Merlin 32 with +18lb max boost; three bladed de Havilland or Rotol propeller.

Dimensions: Wing span 36ft 10in (11.23m); length 29ft 11in (9.12m); height 11ft 5in (3.48m); wing area 242sq ft (22.5m²).

Weights: PR.IV – empty 4,953lb (2,246kg); max loaded 7,200lb (3,266kg). PR.VII – empty 4,985lb (2,261kg); loaded 6,584lb (2,986kg). PR.XIII – empty 5,085lb (2,307kg); max loaded 6,500lb (2,948kg).

Armament: Usually none, see notes.

Performance: PR.IV – max speed 372mph (598km/h); max range 2,000 miles (3,218km).

PR.XIII – max speed 345mph (555km/h) at 10,000ft (3,050m), 332mph (534km/h) at 2,000ft (610m), 327mph (526km/h) at 24,000ft (7,315m); max climb 3,780ft (1,152m)/min; time to 18,000ft (5,485m) 4.9min; service ceiling 37,000ft (11,280m); range 700 miles (1,126km).

Production: New build – 2 Mk.I(PR), 229 PR.IV, 15 PR Type F (PR.IV), all by Supermarine; also various conversions (see notes).

Notes: Photo-reconnaissance Spitfires played a significant role in 1939-45 and beyond and were built in numerous versions both by conversion and from new production. The first two conversions from Spitfire Is were made in late 1939 (see *Birth of the PR Spitfire* breakout box). The early model PR Spitfires were given alphabetical designations initially and numerical designations later on, some of them official and others not. Those based on the Mk.I were sometimes collectively designated as Spitfire PR.IIIs but other designations were also applied. The various early model PR Spitfires are summarised below.

PR Type A (PR.IA): Simple Spitfire I conversion with standard 85imp gal (386 l) fuel capacity and two F.24 cameras under wings. Applies to the first two conversions.

PR Type B (PR.IB): Similar to Type A but with additional fuel tank in rear fuselage bringing total capacity 114imp gal (518 l); first delivery January 1940.

PR Type C (PR.IC): Two F.24 cameras in blister under starboard wing and one rear fuselage camera; additional fuel

Spitfire PR Type F (X4492) converted from a Mk.I.

tank under port wing, total capacity 144imp gal (655 l); first delivery March 1940, 15 conversions by Heston Aircraft and RAE Farnborough; sometimes designated PR.III.

PR Type D (PR.ID): Extra long range version with redesigned wing internal structure incorporating additional 66½imp gal (302 l) in each leading edge to give total capacity of 218imp gal (991 l); additional engine oil in blister under port wing; two F.8 or F.24 cameras in rear fuselage to give overlapping image; formed basis of new production PR.IV and converted Type Ds later carried same designation.

PR Type E (PR.IE): Single conversion (N3117) with fuel capacity as Type B and one oblique F.24 camera under each wing; optimised for low altitude photography; sometimes designated PR.V.

PR Type F (PR.IF): Underwing and rear fuselage fuel tanks, total capacity 174imp gal (791 l); deepened lower cowling with additional oil capacity; two vertical F.24 cameras in fuselage; conversions by Heston Aircraft in 1941-42 from new production by Supermarine; sometimes designated PR.VI and also sometimes included in PR.IV designation.

PR Type G (redesignated PR.VII): Conversion of Spitfire VA by Heston Aircraft with eight 0.303in machine guns retained; rear fuselage fuel tank giving 114imp gal (518 l) capacity; one oblique and two vertical F.24 cameras in rear fuselage.

PR.IV: New build version of PR Type D based on Spitfire V with Merlin 45/50/55/56 engines. Three camera installation options: 'W' fitting – two F.8 (20in lens); 'X' fitting – one F.24 (14in lens); or 'Y' fitting – one F.52 (36in lens).

PR.XIII: 18 conversions of Spitfire Is, IIs, Vs and PR.VIIs (first delivery April 1943) optimised for low level PR with low blown Merlin 32, standard internal fuel capacity (plus optional overload tanks) and one oblique and two vertical F.24 cameras.

Spitfire PR Type E N3117 (left) began life as a Mk.I and then became a PR Type C before assuming its final configuration. PR.IV BR416 (right) was built from new as a photo-reconnaissance aircraft. (via Neil Mackenzie)

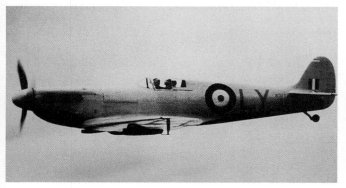

Standard Spitfire VBs of 243 Squadron RAF with Castle Bromwich built EN821 in the foreground. (via Neil Mackenzie)

Type numbers: Mk.VA/B – 331. Mk.VC – 349. Mk.VB (Tropicalised) – 352.

First flight: 26 December 1940 (X4942 converted from Mk.I).

Service entry: February 1941.

Powerplant: VA/B/C – 1,470hp (1,096kW) Merlin 45, 50, 55 or 56 with 1-speed/1-stage supercharger and +16lb max boost. LF.V – 1,585hp (1,182kW) Merlin 45M, 50M or 55M with +18lb max boost; three bladed Rotol or de Havilland propeller. Internal fuel 85imp gal (386 l) in forward fuselage tanks or 114imp gal (518 l) including rear fuselage tank (used only with large drop tank); provision for 30imp gal (136 l), 90imp gal (409 l) or 170imp gal (773 l) underfuselage drop tanks.

Dimensions: Wing span 36ft 10in (11.23m); length 29ft 11in (9.12m); height 11ft 5in (3.48m); wing area 242sq ft (22.5m²). Clipped wing – span 32ft 6in (9.91m); area 231sq ft (21.5m²).

Weights: VA – empty 4,981lb (2,259kg); normal loaded 6,416lb (2,910kg); max loaded 6,700lb (3,039kg). VB – empty 5,065lb (2,297kg); normal loaded 6,622lb (3,004kg); max loaded 6,700lb (3,039kg) VC – empty 5,081lb (2,305kg); normal loaded 6,785lb (3,078kg); max loaded 7,420lb (3,366kg).

Armament: VA – eight 0.303in Browning machine guns within wings with 350rpg. VB – two 20mm Hispano cannon (120rpg) and four 0.303in Browning machine guns (350rpg). VC – as VA/B or four 20mm Hispano cannon with 120rpg; provision for one 500lb (227kg) or two 250lb (113kg) bombs.

Performance: F.V – max speed 371mph (597km/h) at 20,000ft (6,095m), 331mph (533km/h) at 10,000ft (3,050m); economical cruise 226mph (364km/h); max climb 3,250ft (990m)/min; time to 20,000ft (6,095m) 7.4min; service ceiling 37,000ft (11,277m); normal range 470 miles (756km); range with max external fuel 1,170 miles (1,883km).
LF.V – max speed 357mph (574km/h) at 6,000ft (1,830m); economical cruise 195mph (314km/h); max climb 4,720ft (1,438m)/min; time to 8,000ft (2,430m) 1.75min; time to 18,000ft (5,485m) 4.8min; service ceiling 36,500ft (11,125m); normal range 475 miles (764km); range with drop tanks 990 miles (1593km).

Tropical filter variations: Spitfire VC EE610 (left) with the bulbous Vokes unit; and Mk.VC EP688 of 40 Squadron SAAF with the more compact Aboukir filter. (via Neil Mackenzie)

Mk.VC BR202 carrying the 170imp gal (773 l) overload tank used when aircraft were ferried from Britain to the Mediterranean area.

Production: 94 Mk.VA, 3,911 Mk.VB, 2,467 Mk.VC, total 6,472 (plus 15 completed as PR Type F) by Supermarine (1,363), CBAF (4,489) and Westland (635).

Notes: The Mk.V was the most produced of all the Spitfire variants and the first to be widely deployed, serving the RAF and its allies over Europe, Malta, the Western Desert, Australia, India-Burma, the Far and Middle East, Mediterranean and Eastern Front. It was also the first to be developed into a substantial family of variants and the first to be operated in numbers by foreign air forces.

Utilising the basic airframe of the Mks.I and II in combination with the more powerful Merlin 45 engine (which reverted to electrical starting rather than the Mk.II's cartridge system), the first Mk.V (converted from a Mk.I) was flown in December 1940 and deliveries began in February 1941, initially of aircraft converted from Mk.Is both on the production line and from existing aircraft. New build Mk.Vs began appearing in March 1941 initially from the parent company and Westland, and from CBAF three months later.

Three basic models were built: the Mk.VA with machine guns, Mk.VB with cannon and machine guns and F.VC with the newly developed 'universal' C wing capable of accepting both types of armament or four 20mm cannon. The VC also featured strengthened undercarriage while all versions had a further increase in armour protection and machine gun ammunition increased from 300 to 350 rounds per gun.

The 'C' wing also allowed an increase in cannon ammunition (from 60 to 120 rounds per gun) thanks to a new belt feed system in place of the earlier drums. The VC first appeared in October 1941. The RAF applied role designation prefixes from 1942, resulting in the designations F.VA, F.VB and F.VC.

Ongoing improvements saw metal (instead of fabric covered) ailerons fitted as well as Merlin 50/55 engines with negative 'g' carburettor, improved armour protection and numerous other equipment and specification modifications including pressurised fixed fuel tanks, structural strengthening, negative 'g' carburettor (where it was not already fitted), additional armour, fuselage flush riveting and many more.

Drop tanks were also developed for the Mk.V, as were various bomb fits and tropicalised versions from early 1942, most with the large Vokes filter under the nose. Other tropicalised Spitfire Vs had the more compact Aboukir filter,

developed by an Egypt based Maintenance Unit. Production of the Mk.VC continued until July 1943.

From 1943 most Spitfire Vs were converted to LF.VA/B/Cs with Merlin 45M, 50M or 55M engines with cropped supercharger impellers and optimised for maximum power output at low altitudes. Clipped wings were also fitted to improve manoeuvrability at the low altitudes at which most of the aircraft were by now operating.

Although outclassed as a fighter by the Focke-Wulf Fw 190 (its arrival necessitating rapid development of the Spitfire IX) the Mk.V was the first multirole Spitfire variant and one which became a versatile workhorse of the mid war years on all fronts.

Spitfire VCs with rare four cannon armament.

Spitfire VI

The prototype Spitfire VI (X4942), the first model designed specifically for high altitude operations. Features included a lightly pressurised cabin, a high rated Merlin 46 and – as is well shown here – extended span wings. X4942 was converted from a Mk.V.

Type number: 350.

First flight: 4 July 1941 (X4942 converted from Mk.V).

Service entry: April 1942.

Powerplant: 1,415hp (1,055kW) Rolls-Royce Merlin 47 with 1-speed/1-stage supercharger and +16lb max boost; Rotol four bladed propeller. Internal fuel 85imp gal (386 l) in forward fuselage tanks; provision for 30imp gal (136 l) or 90imp gal (409 l) underfuselage drop tanks.

Dimensions: Wing span 40ft 2in (12.24m); length 29ft 11in (9.12m); height 11ft 5in (3.48m); wing area 249sq ft (23.1m²).

Weights: Empty 5,300lb (2,404kg); normal loaded 6,797lb (3,083kg); max loaded 7,178lb (3,256kg).

Armament: Two 20mm Hispano cannon (60rpg) and four 0.303in Browning machine guns (350rpg).

Performance: Max speed 364mph (586km/h) at 22,000ft (6,705m), 264mph (425km/h) at 38,000ft (11,580m); max cruise 325mph (523km/h) at 20,000ft (6,095m); normal cruise 239mph (385km/h); time to 20,000ft (6,095m) 8.7min; time to 38,000ft (11,580m) 27.9min; service ceiling 39,200ft (11,948m); range (internal fuel) 510 miles (820km); max range with external fuel 1,170 miles (1,883km).

Production: 100 by Supermarine.

Notes: The threat of high flying German bombers and reconnaissance aircraft operating over Britain from 1941 resulted in the urgent necessity to develop Spitfire variants capable of intercepting these aircraft at altitudes beyond the normal reach of other models.

By then, Rolls-Royce had begun developing the art of producing Merlin variants in which the supercharger ratios could be 'tuned' to suit particular applications, such as was being done with the low altitude rated versions for the Spitfire LF.V.

This customising of power outputs would subsequently be fully exploited when the Merlin 60 and 70 engines with two-speed/two-stage superchargers were introduced for later model Spitfires, in conjunction with greater levels of boost and higher octane fuels.

Based on the Mk.VB wing and airframe, the Spitfire VI (sometimes HF.VI, the 'H' for 'high') featured the Merlin 47 which was designed to produce its maximum output at 14,000ft (4,267m).

It also incorporated other features for its high altitude role including a four bladed propeller (the first Spitfire production variant to have this), extended wingtips for better controllability at height and a lightly pressurised (2.0psi) cabin resulting from research undertaken by the Royal Aircraft Establishment and Supermarine during 1940-41.

The pressure cabin was housed between bulkheads fore and aft of the cabin and the sealed cockpit canopy could not be slid open – it was bolted down and sealed once the pilot was in the cockpit but could be jettisoned in its entirety in an emergency.

The Merlin 47 featured an extra accessory drive to which was attached a Marshall blower to force air into the cockpit (via vents which were set up to allow them to double as canopy demisters), this sealed within the pressure cabin. Controls runs, cables, pipes and other items which had to pass through the pressure cabin were provided with airtight seals on the bulkheads.

The prototype (converted from a Mk.V which in turn had begun life as a Mk.I) first flew in July 1941 and the Mk.VI entered service in April 1942 with 616 Squadron.

The first action was against a Dornier Do 217 the following month. Production of the Spitfire VI was limited to just 100 aircraft (the last delivered in November 1942) and five home based RAF squadrons operated it during 1942-43 with only limited success as performance was only fair. Unable to fight at altitudes above about 35,000ft (10,670m), the aircraft was no match for the high altitude Junkers Ju 86P and R bomber-reconnaissance aircraft which was its main adversary.

The sealed canopy was also unpopular with pilots as normal procedure was to taxy, takeoff and land with the canopy open. Efficient heating meant the aircraft could be flown in shirtsleeves at high altitudes, but this kind of dress was not practical in combat with the result that pilots were often 'cooked' when wearing more usual kit. A bypass system was duly devised.

The Spitfire VI was regarded as an interim high altitude interceptor pending availability of the much more capable Mk.VII with two-speed/two-stage Merlin.

The Spitfire VII entered regular service in March 1943.

Type number: 351.

First flight: 22 August 1942 (AB450).

Service entry: March 1943.

Powerplant: F.VII – 1,565hp (1,167kW) Rolls-Royce Merlin 61 with 2-speed/2-stage supercharger and +15lb max boost or 1,710hp (1,275kW) Merlin 64 with +18lb max boost. HF.VII – 1,655hp (1,234kW) Merlin 71 (+18lb max boost); four bladed Rotol propeller. Internal fuel 122imp gal (555 l) in forward fuselage and wing leading edge tanks; provision for 30imp gal (136 l), 90imp gal (409 l) or 170imp gal (773 l) underfuselage drop tank.

Dimensions: Wing span 40ft 2in (12.24m); length 31ft 3½in (9.54m); height 12ft 7¾in (3.86m); wing area 249sq ft (23.1m²).

Weights: Empty 6,000lb (2,722kg); loaded 7,928lb (3,596kg).

Armament: Two 20mm Hispano cannon with 120rpg and four 0.303in Browning machine guns with 350rpg.

Performance: F.VII – max speed 408mph (656km/h) at 25,000ft (7,620m), 382mph (615km/h) at 12,500ft (3,810m); max cruise 324mph (521km/h) at 20,000ft (6,095m); economical cruise 220mph (354km/h); time to 20,000ft (6,095m) 7.1min; service ceiling 43,000ft (13,106m); normal range 660 miles (1,062km).
HF.VII – max speed 424mph (682km/h) at 29,400ft (8,960m), 400mph (644km/h) at 20,000ft (6,095m), 409mph (658km/h) at 40,000ft (12,190m); max climb 4,060ft (1,237m)/min; time to 20,000ft (6,095m) 5.4min; time to 40,000ft (12,190m) 15.5min; service ceiling 45,100ft (13,745m).

Production: 140 by Supermarine.

Notes: Although featuring earlier mark numbers, the Spitfire VII (and VIII) followed the Mk.IX into service and represented the first substantial redesign of the basic aircraft to achieve production. Both had the 60 series two-speed/ two-stage Merlins, six port exhausts, four bladed propeller, symmetrical underwing radiators and longer nose of the Mk.IX in combination with a structurally modified and stronger fuselage with a revised 'C' wing incorporating shorter span ailerons and extra fuel tanks in the leading edges.

A retractable tail wheel was fitted as was – in most cases – the 'pointed' fin and rudder of increased area which became necessary due to the destabilising effect of the new propeller's high solidity.

The Spitfire VII was a specialised high altitude interceptor model with extended span wings and a pressurised cabin fed by a Marshall Mk.XII blower. Most of the only 140 built were fitted with the Merlin 64 engine and these (plus a few with Merlin 61s) were designated as the F.VII.

A small number had the Merlin 71 installed and carried the designation HF.VII, these demonstrating superb high altitude performance but by the time they entered service the threat of very high flying German raiders – the reason for which the Spitfire VII was developed – had passed. Only 16 Merlin 71 engines were manufactured.

The first Mk.VII flew in April 1942 and production aircraft began rolling of the line the following September. The last example was delivered in May 1944. No 124 Squadron was the first of only six operational units to put the aircraft into service, in March 1943. When operating at extreme altitudes, the Spitfire VII often flew with its armour, cannon and wing tanks removed in order to save over 900lb (408kg) of weight to help squeeze a bit more performance.

Spitfire LF.VIII JG204, used as the tropical trials aircraft.

Type number: Prototype – 359. Production 360.

First flight: 20 November 1942 (JF274).

Service entry: August 1943.

Powerplant: F.VIII – 1,565hp (1,167kW) Merlin 61 with 2-speed/2-stage supercharger and +15lb max boost or 1,710hp (1,275kW) Merlin 63/63A (+18lb max boost). LF.VIII – 1,720hp (1,282kW) Merlin 66 (+18lb max boost). HF.VIII – 1,710hp (1,275kW) Merlin 70 (+18lb max boost); four bladed Rotol propeller. Internal fuel 122imp gal (555 l) in forward fuselage and wing leading edge tanks; provision for 30imp gal (136 l), 90imp gal (409 l) or 170imp gal (773 l) drop tank.

Dimensions: Wing span 36ft 10in (11.23m); length 31ft 3½in (9.54m); height 12ft 7¾in (3.85m); wing area 242sq ft (22.5m²). Some HF.VIIIs with extended wing of 40ft 2in (12.24m) span and area 249sq ft (23.1m²).

Weights: Typical empty 5,805lb (2,633kg); normal loaded 7,807lb (3,541kg); max loaded 8,021lb (3,638kg).

Armament: Two 20mm Hispano cannon with 120rpg and four 0.303in Browning machine guns with 350rpg; or four 20mm Hispano cannon with 120rpg; provision for one 500lb (227kg) and two 250lb (113kg) bombs.

Performance: F.VIII – max speed 408mph (656km/h) at 25,000ft (7,620m), 382mph (615km/h) at 12,500ft (3,810m); max cruise 324mph (521km/h) at 20,000ft (6,095m); time to

RAAF Spitfire LF.VIIIs with A58-402 (ex MD232) in foreground.

20,000ft (6,095m) 7.0min; service ceiling 43,000ft (13,105m); normal range 660 miles (1,062km); max range with drop tank 1,180 miles (1,900km).

LF.VIII – max speed 404mph (650km/h) at 21,000ft (6,400m), 384mph (618km/h) at 10,500ft (3,200m), 362mph (582km/h) at sea level; max climb 4,800ft (1,463m)/min; time to 20,000ft (6,095m) 5.7min; service ceiling 41,500ft (12,650m).

HF.VIII – max speed 416mph (669km/h) at 27,000ft (8,230m); service ceiling 44,000ft (13,410m).

Production: 267 F.VIII, 1,225 LF.VIII and 160 HF.VIII, total 1,652 by Supermarine (plus 6 completed as Mk.XIV prototypes).

Notes: The Spitfire VIII was structurally and aerodynamically similar to the Mk.VII but without the pressure cabin and intended for fighter and fighter-bomber roles. It incorporated the new features of the Mk.VII including the modified structure, reduced span ailerons, retractable tailwheel, symmetrical underwing radiators, wing fuel tanks, strengthened undercarriage, six port exhaust and 60 series Merlin with two-speed/two-stage supercharger housed in a lengthened nose and driving a four bladed propeller.

As it was intended mainly for service in the Near East, Far East and Mediterranean, all Mk.VIIIs were tropicalised on production, incorporating a compact Vokes Aero-Vee filter on the carburettor intake under the nose. Three subvariants were built: the 'standard' F.VIII, high altitude HF.VIII (some with extended wingtips) and the major production low altitude LF.VIII, a small number of which had clipped rather than standard span wings.

Other features included the fitting of 'C' armament, and on all but some of the earliest examples, the pointed fin and rudder of increased area. One Mk.VIII – JF299 – was the first Spitfire with the 'rear vision' bubble canopy, appearing in this guise in September 1943. Six were completed as prototypes for the Griffon engined Spitfire XIV.

The Mk.VIII first flew in November 1942 and production began immediately, initially at a slow rate. The first RAF squadron to re-equip was number 145 in Italy in August 1943. Australia was a major recipient of the type for service in the Pacific, receiving 410 from October 1943.

Spitfire Floatplanes

The sole Spitfire IX floatplane conversion (MJ892, right) first flew in May 1944. W3760 (bottom left and right) was the first of the three Mk.V conversions performed in 1942-43.

Type numbers: V – 355. IX – 385.

First flight: V – September 1942 (W3760). IX – May 1944 (MJ892).

Service entry: Not applicable.

Powerplant: V – 1,470hp (1,096kW) Rolls-Royce Merlin 45 with 1-speed/1-stage supercharger and +16lb max boost; four bladed Rotol propeller. Internal fuel 85imp gal (386 l).

Dimensions: V – wing span 36ft 10in (11.23m); overall length 35ft 4in (10.77m); wing area 242sq ft (22.5m²).

Weights: V – empty 6,014lb (2,728kg); loaded 7,580lb (3,438kg).

Armament: Two 20mm Hispano cannon with 120rpg and four 0.303in Browning machine guns with 350rpg.

Performance: V – max speed 324mph (521km/h) at 19,500ft (5,945m), 251mph (404km/h) at sea level, 285mph (458km/h) at 10,000ft (3,050m); max climb 2,450ft (747m)/min; time to 10,000ft (3,050m) 4.1min; time to 20,000ft (6,095m) 9.1min; service ceiling 33,400ft (10,180m).

Production: 3 Mk.V and 1 Mk.IX conversions (plus 1 unflown Mk.I conversion).

Notes: Investigations into the potential of a Spitfire (and Hawker Hurricane) floatplane first emerged in April 1940 during the Norwegian campaign when the RAF's operations were hampered by a lack of airfields. The idea was to operate the aircraft from sheltered waters when land facilities were scarce.

Spitfire I R6722 was converted to take a pair of Blackburn Roc floats. The concept was satisfactorily tank tested but the end of the Norwegian campaign – and the urgent need for landplane Spitfires at the time – saw the requirement pass.

R6722 never flew as a floatplane and was reconverted to standard configuration.

The idea lay dormant for 18 months until early 1942 when Japan had entered the war and was making extensive use of floatplane fighters in the Pacific. Three Spitfire V conversions were therefore produced, incorporating new floats designed by Supermarine's Arthur Shirvall (who had designed the Schneider Trophy racers' floats) and built by Folland Aircraft. The floats were 25ft 7in (7.80m) long and mounted on cantilever struts.

Other modifications (after initial flight testing) included extending the leading edge of the fin to increase its area while a ventral fin extension was fitted below the rear fuselage. A four bladed propeller was installed and the first conversion (W3760) originally had a Vokes tropical filter installed. W3760 first flew in September 1942, initial flight trials demonstrating good air and water handling characteristics plus satisfactory performance. No armament was fitted initially, but this was subsequently installed.

Two other MK.V floatplanes were converted by Folland (EP751 and EP754) in 1943 and at one time there were plans to produce a further 10. The three Mk.Vs were shipped to Alexandria in October 1943 with the intention of operating against *Luftwaffe* Ju 52 transports flying between Greece and Crete but they were sent home before any operations could be flown as the situation there deteriorated for the Allies.

There were also plans for Folland to produce Spitfire IX floatplane conversions for use in the Pacific. A prototype conversion (Mk.IXB MJ892) flew in May 1944 but the whole concept of Spitfire floatplanes was dropped shortly afterwards.

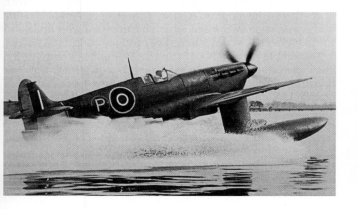

For a 'temporary expedient' the Spitfire IX and closely related Mk.XVI were produced in very substantial numbers. There's one of each in this shot.

Type number: 361.

First flights: F.IX – 29 September 1941 (N3927, converted from Mk.III). LF.IX – 23 May 1943 (MA648, converted from Mk.VC). XVI – December 1943 (MJ556, converted from LF.IX).

Service entry: IX – June 1942. XVI – October 1944.

Powerplant: F.IX – 1,565hp (1,167kW) Rolls-Royce Merlin 61 with 2-speed/2-stage supercharger and +15lb max boost or 1,710hp (1,275kW) Merlin 63/63A (+18lb max boost). HF.IX – 1,710hp (1,275kW) Merlin 70 (+18lb max boost). LF.IX/XVI – 1,720hp (1,282kW) Merlin 66/Packard Merlin 266 (+18lb max boost); four bladed Rotol propeller. Internal fuel (early IX) 85imp gal (386 l) in forward fuselage tanks; (later IX and XVI) up to 126imp gal (573 l) with additional rear fuselage tank (standard canopy) or 118imp gal (536 l) with bubble canopy and modified rear fuselage; additional 36imp gal (164 l) in wing tanks in late models; provision for 30imp gal (136 l), 45imp gal (205 l) or 90imp gal (409 l) drop tank.

Dimensions: Wing span 36ft 10in (11.23m) standard or 32ft 7in (9.93m) clipped; length 31ft 4in (9.55m); height 12ft 7¾in (3.86m); wing area 242sq ft (22.5m²) or 231sq ft (21.5m²).

Weights: Empty 5,800-6,200lb (2,630-2,812kg); normal loaded 7,500lb (3,402kg); max overload 9,500lb (4,309kg).

Armament: IXC/XVI – two 20mm Hispano cannon with 120rpg and four 0.303in Browning machine guns with 350rpg, or four 20mm Hispano cannon with 120rpg. IXE/XVIE – two 20mm Hispano cannon with 350rpg and two 0.50in Browning machine guns with 250rpg; max bomb load 1,000lb (454kg).

Performance: F.IX – max speed 408mph (657km/h) at 25,000ft (7,620m), 326mph (525km/h) at 4,000ft (1,220m); max cruise 324mph (521km/h) at 20,000ft (6,095m); range cruise 220mph (354km/h); max climb 4,100ft (1,250m)/min; time to 20,000ft (6,095m) 5.7min; service ceiling 43,000ft (13,106m); normal range 434 miles (698km); max range with drop tank 980 miles (1,577km).

LF.IX/XVI – max speed 404mph (650km/h) at 21,000ft (6,400m), 384mph (618km/h) at 10,500ft (3,200m); time to 20,000ft (6,095m) 6.4min; service ceiling 42,500ft (12,955m). HF.IX – max speed 416mph (669km/h) at 27,500ft (8,380m), 396mph (637km/h) at 15,000ft (4,570m); max climb 4,470ft (1,362m)/min; time to 30,000ft (9,145m) 8.3min; service ceiling 45,000ft (13,715m).

Production: 4,010 LF.IX, 1,255 F.IX and 400 HF.IX, total 5,665 by Supermarine (561) and CBAF (5,104); and 1,054 LF.XVI by CBAF.

Notes: The appearance of the Focke-Wulf Fw 190 over Europe in 1941 rendered the Spitfire V outclassed as a dogfighter and resulted in the urgent need to develop an upgraded version. Happily, this need coincided with Rolls-Royce developing its more powerful 60 series Merlins with two-speed/two-stage superchargers and the ability to run at higher boosts and therefore produce additional power.

The simple expedient of fitting these engines to a modified Mk.V airframe resulted in the much faster Mk.IX, the first Spitfire to enter service with this new generation of powerplant and ahead of the more substantially redesigned Mks.VII and VIII. Intended as an interim measure, the Spitfire IX went on to be produced in very large numbers and flown by 90 RAF and Commonwealth squadrons. The Spitfire XVI was basically a late model Mk.IX (from October 1944) with a US built Packard Merlin 266 engine, the similarity of the two marks reflected by their shared Supermarine type number.

Rolls-Royce's ability to 'tailor' the engine power of the

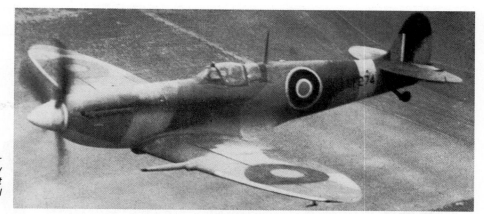

Like many early Spitfire IXs, BS274 was converted from a Mk.V and first flew in its new form in August 1942. Of interest is the fact that in this photograph, the incorrect serial BF274 has been painted on the aircraft.

Late production Spitfire LF.XVIE (RW396) while in service with the Central Gunnery School between 1946 and 1949.

Spitfire IXCs of 601 Squadron RAF while based in Italy during 1944.

Merlin through its supercharger gear ratios was exploited in the Spitfire IX, resulting in three basic models: the 'standard' F.IX for medium altitude work; the low altitude LF.IX and the high altitude HF.IX. Most were LF versions as the need for more ground attack sorties grew as the war progressed. Within those models were the various wing/armament combinations introduced with the Mk.V – IXA with eight machine guns, IXB with cannon and machine guns, and the universal wing IXC – plus from 1944 the IXE with two cannon and two large calibre 0.50in machine guns.

Many were fitted with clipped wings and ongoing improvements included a revised 'pointed' rudder and from late in the production run (December 1944), a cut down rear fuselage and bubble canopy. Internal fuel capacity increased in later models with the introduction of rear fuselage and wing tanks.

All Mk.XVIs were built as LF.XVIs with low/medium rated Packard Merlin 266 engines. The vast majority had clipped 'E' wings and most featured the bubble canopy.

The prototype Mk.IX was converted from the first of the two experimental Mk.IIIs fitted with a Merlin 61, four bladed propeller and standard span wings. This first flew in September 1941 followed by a second (converted from a Mk.I) in January 1942. The Spitfire IX entered squadron service the following June and many early examples were converted from either existing Mk.Vs or aircraft already on the production line.

The first Mk.XVI flew in December 1943 (converted from a Mk.IX) and entry to service was in October 1944.

The Mks.IX and XVI both remained in production throughout the European war, the last Merlin powered Spitfire (an LF.XVIE) delivered in July 1945. A few RAF aircraft were converted to unarmed PR.IX reconnaissance and armed FR.IX tactical reconnaissance models.

Spitfire PR.XI PL775 of 541 Squadron RAF. (via Neil Mackenzie)

Type numbers: PR.X – 362/387. PR.XI – 365/374.

First flight: X – March 1944 (MD192). XI – 21 November 1942 (BS497).

Service entry: X – May 1944. XI – December 1942.

Powerplant: X – 1,710hp (1,275kW) Rolls-Royce Merlin 70 or 1,680hp (1,253kW) Merlin 77 with 2-speed/2-stage supercharger and +18lb max boost. XI – 1,565hp (1,167kW) Merlin 61 (+15lb max boost) or 1,710hp (1,275kW) Merlin 63/63A/70 (+18lb max boost); four bladed Rotol propeller. Internal fuel (X) 228imp gal (1,036 l) in forward fuselage, rear fuselage and wing tanks; (XI) 217imp gal (986 l); provision for 90imp gal (409 l) or 170imp gal (773 l) overload drop tank.

Dimensions: Wing span 36ft 10in (11.23m); length 31ft 10in (9.70m); height 12ft 7¾in (3.85m); wing area 242sq ft (22.5m²).

Weights: Empty 5,812lb (2,636kg); loaded 7,900lb (3,583kg); max overload 8,700lb (3,946kg).

Armament: None.

Performance: XI – max speed 417mph (671km/h) at 24,200ft (7,375m), 387mph (623km/h) at 38,000ft (11,580m); max cruise 397mph (639km/h) at 31,000ft (9,450m); max climb 4,350ft (1,326m)/min; time to 20,000ft (6,095m) 5.0min; service ceiling 44,000ft (13,410m); radius of action with 90imp gal (409 l) drop tank 565 miles (909km); max range with 170imp gal (773 l) overload tank 2,300 miles (3,700km).

Production: 16 PR.X and 471 PR.XI by Supermarine.

Notes: The PR.X and PR.XI represented the second generation of Spitfire photo-reconnaissance variants, powered by 60 or 70 series Merlins and featuring the airframe characteristics of the Mks.IX, VII and VIII fighters, including the retractable tailwheel of the latter pair in all but the earliest PR.XIs.

The unpressurised PR.XI appeared first, first flying in November 1942 and entering service shortly afterwards. Such was the urgent need for such a PR Spitfire, the Photographic Reconnaissance Unit (PRU) at Benson converted 15 Mk.IX fighters to PR.IXs with two cameras in the rear fuselage. These lacked the PR.XI's wing tanks.

Production PR.XIs were built to two standards: early models converted from Mk.IX airframes (Type 374) with Merlin 61 or 63 engines and the Mk.IX's fixed tailwheel and original shape fin and rudder; and subsequent aircraft to the definitive standard (Type 365) with Merlin 70, retractable tailwheel and pointed fin and rudder. All had standard span wings and some later aircraft had Merlin 77 engines.

The standard camera options were similar to the PR.IV – two vertical and one oblique F.24 or two vertical F.24s or F.52s. Later aircraft had a universal mounting system which allowed any configuration to be quickly loaded. The PR.XI became the major photo-reconnaissance Spitfire variant from 1943 with the last of the 471 built delivered in February 1945.

The pressurised PR.X appeared considerably later, not entering service until 1944 due to it being given a low priority. Only 16 were built, the aircraft incorporating the pressure cabin features of the Spitfire VII but otherwise similar to the PR.XI. It wasn't very popular with pilots and remained in service for little over a year with just two squadrons. Operationally, it was realised that pressurisation wasn't necessary for most sorties, visibility was poor through the PR.X's thick canopy and there was the added maintenance and serviceability burden created by the pressure cabin's systems.

Spitfire T.8 and T.9

The sole Spitfire T.8 carrying the Class B registration N32.

Type numbers: T.8 – 502. T.9 – 509.

First flights: T.8 – May 1946 (N32/G-AIDN). T.9 – 1948 (H-97 ex MK715).

Service entry: T.8 – not applicable. T.9 – March 1948.

Powerplant: 1,720hp (1,282kW) Rolls-Royce Merlin 66 with 2-speed/2-stage supercharger and +18lb max boost; four bladed Rotol propeller. Internal fuel 96imp gal (436 l) in forward fuselage and wing tanks; provision for 30imp gal (136 l) drop tank.

Dimensions: Wing span 36ft 10in (11.23m); length 31ft 4$\frac{1}{2}$in (9.56m); height 12ft 7$\frac{3}{4}$in (3.85m); wing area 242sq ft (22.5m^2).

Weights: T.8 – empty 5,950lb (2,699kg); loaded 7,400lb (3,357kg). T.9 – empty 5,850lb (2,654kg); loaded 7,300lb (3,311kg).

Armament: Two 20mm Hispano cannon or four 0.303in Browning machine guns optional.

Performance: T.8 – max speed 393mph (632km/h) at 20,000ft (6,095m), 326mph (525km/h) at sea level; range cruise 232mph (373km/h); max climb 4,570ft (1,393m)/min; service ceiling 40,600ft (12,375m); normal range 240 miles (386km); range with drop tank 418 miles (673km).
T.9 – max speed 386mph (621km/h) at 20,000ft (6,095m), 322mph (518km/h) at sea level; max climb 3,970ft (1,210m)/min; service ceiling 41,500ft (12,650m); normal range 234 miles (376km); range with drop tank 418 miles (672km).

Production: 1 T.8 conversion and 20 T.9 conversions by Supermarine.

Notes: Even though Supermarine had proposed the concept of a two seat trainer version of the Spitfire as early 1941, official interest was not forthcoming and the idea remained in the background until after the war. There had been some unofficial conversions in the meantime, however, No 261 Squadron based in Sicily converting one of its Spitfire VCs (ES127) into a two seater in 1944 and the Soviets also converting several Spitfire IXs.

The first 'official' Spitfire Trainer was converted as a private venture from LF.VIII MT818 which had originally flown in May 1944 and been used for trials. It was purchased by Vickers-Armstrong and converted to a Type 502 Spitfire Trainer, first flying in its new guise in May 1946 and carrying the Class B registration N32. It was subsequently registered G-AIDN and acquired the unofficial designation Spitfire T.8 (or T.VIII and sometimes Tr.8).

The conversion involved moving the front cockpit 13.5in (34.3cm) forward and inserting a second, raised cockpit for an instructor behind. Moving the front cockpit resulted in the fuselage fuel tank's capacity being more than halved, necessitating the installation of tanks in the gun bays in addition to the normal leading edge tanks. G-AIDN was demonstrated widely, including to the RAF in April 1949 to fill a possible Royal Auxiliary Air Force requirement for about 20 aircraft, but no orders were forthcoming and the T.8 remained a one-off.

Supermarine had more success with the Type 509, converted from Mk.IX Spitfires and unofficially designated T.9 (or T.IX/Tr.9). This conversion was similar to the T.8 but had a fixed tailwheel and the additional wing fuel tanks located between the wing spars.

The Netherlands was the first customer, ordering three (ex MK715, BF274 and BS147) for delivery in March 1948. Other orders were received from India (10), Egypt (1) and Ireland (6), the latter's the last of the line and delivered between May and July 1951. Supermarine also studied a trainer version of the Spitfire XVIII as the Type 518 but this was not built.

Type number: 366.

First flight: 24 August 1942 (DP845, converted from Mk.IV/XX).

Service entry: February 1943.

Powerplant: One 1,735hp (1,294kW) Rolls-Royce Griffon III or IV with 2-speed/1-stage supercharger and +12lb max boost; four bladed Rotol propeller. Internal fuel 85imp gal (386 l); provision for 30imp gal (136 l) underfuselage drop tank.

Dimensions: Wing span 32ft 7in (9.93m); length 31ft 10in (9.70m); height 11ft 0in (3.35m); wing area 231sq ft (21.5m²).

Weights: Empty 5,600lb (2,540kg); loaded 7,400lb (3,357kg).

Armament: Two 20mm Hispano cannon with 120rpg and four 0.303in Browning machine guns with 350rpg; provision for one 500lb (227kg) or two 250lb (113kg) bombs.

Performance: Max speed 393mph (632km/h) at 18,000ft (5,486m), 372mph (598km/h) at 5,500ft (1,675m); 346mph (557km/h) at sea level; range cruise 263mph (423km/h); max climb 3,800ft (1,158m)/min; time to 20,000ft (6,095m) 6.7min; service ceiling 40,000ft (12,192m); normal range 329 miles (529km); range with drop tank 493 miles (793km).

Production: 100 by Supermarine.

Notes: The quest for more performance in the Spitfire led to installing the Rolls-Royce Griffon engine in the aircraft. With a one-third greater capacity than the Merlin, the Griffon promised proportionately higher power outputs.

After some 18 months' development work by Supermarine, the Spitfire's formal association with the Griffon began in early 1940 when two prototypes were ordered. Powered by a 1,735hp (1,294kW) Griffon IIB with two-speed/single-stage supercharger driving a four bladed propeller and designated Mk.IV (later changed to Mk.XX to avoid confusion with the PR.IV), the first aircraft flew on 27 November 1941. Orders were placed for 750 Mk.XXs but circumstances dictated development of a different version for production

The need for a fighter with high performance at low altitude to counter Fw 190 'hit and run' raids on England's south coast resulted in development of the first production Griffon powered Spitfire, the F.XII with low rated Griffon III or IV engine. After completing trials with a mockup six cannon installation, the first Mk.XX was converted to become the prototype F.XII, flying in this guise for the first time on 24 August 1942. The first production Spitfire XII flew in October 1942.

The Spitfire XII featured the restyled engine cowling necessitated by the larger Griffon (which also had a lower thrust line than the Merlin and rotated in the opposite direction), clipped 'C' wings and pointed rudder. Early aircraft were based on the Mk.V airframe with fixed tailwheel while later examples used the Mk.VIII as their basis and had a retractable tailwheel. In all cases the internal fuel capacity was the same as the Spitfire I/II/V models and the Mk.XII's range was therefore poor. The ability to carry a 30imp gal (136 l) drop tank helped only slightly but for the short range interception duties for which the aircraft was designed, this was not a major operational problem.

The Spitfire XII entered service in February 1943 and the small number built served with two RAF home defence squadrons until 1944. Very fast at low level, it was successful in its intended role, working in conjunction with Typhoon squadrons. On one occasion in May 1943 Spitfire XIIs of 91 Squadron claimed five Fw 190s in a single day. Both squadrons also flew shipping reconnaissance missions and fighter sweep sorties into France.

(above left) The Mk.XII was the first Griffon engined Spitfire to enter production. It was built in small numbers and served with only two RAF front line squadrons. (left) Spitfire XII MB882 of 41 Squadron. (via Neil Mackenzie).

Spitfire F.XIVE.

Type numbers: Prototypes – 369; Griffon 85 engine – 373; main production – 379.

First flight: 4 March 1943 (JF317, converted from Mk.VIII).

Service entry: January 1944.

Powerplant: 2,050hp (1,528kW) Rolls-Royce Griffon 65 with 2-speed/2-stage supercharger and +18lb max boost; five bladed Rotol propeller. Some with 2,055hp (1,532kW) Griffon 85 (+18lb max boost) and six bladed Rotol contra-props. Internal fuel (F.XIV) 111imp gal (505 l) in forward fuselage and wing leading edge tanks; FR.XIV additional rear fuselage tank, total capacity 142imp gal (645 l); provision for 30imp gal (136 l), 45imp gal (205 l), 50imp gal (227 l) or 90imp gal (409 l) underfuselage drop tank.

Dimensions: F.XIV – wing span 36ft 10in (11.23m); length 32ft 8in (9.96m); height 12ft 8in (3.86m); wing area 242sq ft (22.5m^2). FR.XIV – wing span 32ft 7in (9.93m); wing area 231sq ft (21.5m^2).

Weights: F.XIVE – empty 6,600lb (2,994kg); normal loaded 8,500lb (3,856kg); max overload 10,280lb (4,663kg).

Armament: F.XIV – two 20mm Hispano cannon with 120rpg and four 0.303in Browning machine guns with 350rpg. F/FR.IVE – two 20mm Hispano cannon with 120rpg and two 0.50in Browning machine guns with 250rpg; one 500lb (227kg) or two 250lb (113kg) bombs.

Performance: Max speed 448mph (721km/h) at 26,000ft (7,925m), 375mph (603km/h) at sea level, 417mph (671km/h) at 12,000ft (3,660m); max cruise 362mph (582km/h) at 20,000ft (6,095m); range cruise 245mph (394km/h); max climb 4,700ft (1,432m)/min; time to 20,000ft (6,095m) 5.1min; service ceiling 44,500ft (13,565m); normal range 460 miles (740km); max range with drop tank 850 miles (1,368km).

Production: 527 F.XIV/E and 430 FR.XIVE, total 957 by Supermarine (plus six prototypes converted from Mk.VIIIs).

Notes: Development of the Griffon powered Spitfire continued with the Mk.XIV powered by a Griffon 65 with intercooled two-speed/two-stage supercharger and driving a five bladed propeller. The new engine necessitated a longer nose which in turn required the fitting of a new fin and rudder of broader chord to compensate. Other Mk.XIV features were a retractable tailwheel and the wing fuel tanks and revised ailerons of the Mk.VIII.

Six converted Mk.VIIIs served as prototypes for the Mk.XIV, the first one flying in March 1943. The first true Mk.XIV flew in October 1943 and squadron service began in January 1944.

Three variants were produced: the standard F.XIV with 'C' wing and 'normal' Spitfire canopy; F.XIVE with 'E' wing (some clipped for low level operations) and bubble canopy in late production aircraft; and the FR.XIVE low level fighter/ tactical reconnaissance version with clipped wings, bubble canopy, oblique camera in the rear fuselage and additional rear fuselage fuel tank.

Very fast at all altitudes, manoeuvrable and featuring a prodigious rate of climb, the Spitfire XIV proved itself to be the match of any other fighter and had some success against jets including claiming the first Messerschmitt Me 262 to be shot down, in October 1944. It was also very successful against the V-1 flying bomb (over 300 claimed) and as an attack aircraft flying with the 2nd Tactical Air Force in Europe. More than 30 RAF and Commonwealth squadrons were equipped with the Mk.XIV, despite being officially regarded as merely a 'useful interim type' until the heavily revised (and much delayed) Spitfire 21 entered service.

Spitfire FR.XIV RN218 of 273 Squadron RAF.

Early production Spitfire F.XIV RB146 was used for trials.

Spitfire XVIII

Type number: 394.

First flight: March 1945 (SM843).

Service entry: May 1945.

Powerplant: 2,050hp (1,529kW) Rolls-Royce Griffon 65 or 67 with 2-speed/2-stage supercharger and +18lb max boost or +25lb with 150 octane fuel; five bladed Rotol propeller. Internal fuel (F.XVIII) 175imp gal (795 l) in forward fuselage, wing and rear fuselage tanks; (FR.XVIII) 142imp gal (645 l) in forward fuselage, rear fuselage and wing tanks; provision for 30imp gal (136 l), 45imp gal (204 l), 50imp gal (227 l), 90imp gal (409 l) or 170imp gal (773 l) drop tank.

Dimensions: Wing span 36ft 10in (11.23m); length 33ft 3¼in (10.14m); height 12ft 8in (3.86m); wing area 242sq ft (22.5m²).

Weights: Empty 6,850lb (3,107kg); normal loaded 9,100lb (4,128kg); max overload 11,000lb (4,990kg).

Armament: Two 20mm Hispano cannon with 120rpg and two 0.50in Browning machine guns with 250rpg; three 500lb (227kg) bombs or rocket projectiles.

Performance: Max speed 437mph (703km/h) at 24,500ft (7,470m), 357mph (574km/h) at sea level, 402mph (647km/h) at 11,000ft (3,355m), 416mph (669km/h) at 34,000ft (10,365m); max climb 4,200ft (1,280m)/min; service ceiling 44,500ft (13,565m); normal range (F.XVIII) 700 miles (1,126km), FR.XVIII 580 miles (933km).

Production: 100 F.XVIII and 200 FR.XVIII, total 300 by Supermarine.

Notes: Put into service just too late to see action in WWII, the Spitfire XVIII replaced the Mk.XIV on Supermarine's production line in the first half of 1945 with the first deliveries taking place in May 1945.

Similar in most respects to its predecessor including the powerplant, the XVIII shared the late model XIV's major characteristics including the broad chord fin and rudder and bubble canopy but with several notable changes under the skin.

The fuselage was fractionally longer and strengthened, while the undercarriage was stronger to cope with the increased weights associated with the introduction of additional fuel tanks in the wings and rear fuselage.

The 'E' wing with two 20mm cannon and two 0.50in machine guns was outwardly similar to the XIV's but featured a substantial internal redesign, the structure strengthened by using an extruded section (solid) main spar in place of the previous laminated square section tubular unit.

Two versions were produced: the basic F.XVIII fighter-bomber with guns and bombs or rocket armament; and the tactical reconnaissance FR.XVIII which sacrificed one of the rear fuselage fuel tanks in order to accommodate two vertical and one oblique F.24 cameras or a single F.52 vertical.

The Spitfire XVIII (or F.18/FR.18 post war) was built only in small numbers during 1945, the bulk of production occurring between December 1945 and February 1946 when just over two-thirds of the 300 built were manufactured. It was operated by six RAF squadrons in the Far and Middle East and had been withdrawn from front line service by the end of 1951.

The Mk.18 was responsible for making a little bit of history on 1 January 1951 when aircraft from No 60 Squadron based at Tengah flew the last Spitfire combat sortie against terrorists in the Johore area during the Malayan Emergency.

Spitfire FR.18 TP391 of 208 Squadron RAF based in the Suez Canal Zone in 1948.

Spitfire FR.18 TP265 delivered to the RAF in June 1945.

Spitfire PR.XIX

Type numbers: 389 (unpressurised) and 390 (pressurised).

First flight: April 1944 (RM626).

Service entry: May 1944.

Powerplant: 2,050hp (1,529kW) Rolls-Royce Griffon 65 or 66 with 2-speed/2-stage supercharger and +18lb max boost; five bladed Rotol propeller. Internal fuel (early aircraft) 217imp gal (986 l) or (later) 254imp gal (1,155 l) in forward fuselage, wing leading edge and interspar tanks; provision for 30imp gal (136 l), 45imp gal (204 l), 90imp gal (409 l) or 170imp gal (773 l) drop tank .

Dimensions: Wing span 36ft 10in (11.23m); length 32ft 8in (9.96m); height 12ft 8in (3.86m); wing area 242sq ft (22.5m²).

Weights: Empty 6,550lb (2,971kg); loaded 9,202lb (4,174kg); max overload 10,450lb (4,740kg).

Armament: None.

Performance: Max speed 446mph (718km/h) at 26,000ft (7,925m), 360mph (579km/h) at sea level, 410mph (660km/h) at 12,000ft (3,660m); cruising speed 325-375mph (523-603km/h) at 35,000ft (10,670m); long range cruise 250mph (402km/h); max climb 4,400ft (1,341m)/min; time to 35,000ft (10,670m) 15.5min; service ceiling 42,600ft (12,985m); range with internal fuel 680 miles (1,094km) at high speed cruise and 1,120 miles (1,802km) at range cruise; range with 90imp gal drop tank 1,455 miles (2,341km); range with 170imp gal overload tank 1,550 miles (2,495km).

Production: 225 by Supermarine.

Notes: The final photographic-reconnaissance version of the Spitfire and the last of the line to use the signature elliptical wing, the PR.XIX (PR.19 post war) combined the Mk.XIV's powerplant, undercarriage, tail surfaces and basic fuselage with the PR.X's pressure cabin, the PR.XI's camera installations and a variation of the PR.IV/X/XI 'bowser' wing with leading edge and (on all but the earliest models) additional tanks between the spars. The result was a highly capable PR aircraft which could cruise at high speeds and high altitudes over a useful radius of action.

The PR.XIX was supposed to be a temporary measure pending the arrival of a PR version of Supermarine's new fighter, the Spiteful, but its problems, delays and ultimate failure meant the PR.XIX enjoyed a longer than planned production and service career. It was the final Spitfire variant to serve with the RAF.

In order to expedite service entry, the first 22 PR.XIXs lacked the pressure cabin and were powered by Griffon 65 engines. Deliveries began in May 1944. The remainder were pressurised and powered by the Griffon 66, this differing only in featuring the necessary equipment to operate the cabin blower. The prototype pressure cabin PR.XIX (SW777) first flew in October 1944 and the first production model the following month, these also featuring the additional wing fuel tanks between the spars. The 225th and last PR.XIX was delivered in June 1946.

The Spitfire PR.XIX and PR versions of the de Havilland Mosquito formed the backbone of the RAF's photo-reconnaissance capability until PR variants of the Gloster Meteor and English Electric Canberra jets entered service. The RAF's last Spitfire front line operational sortie was performed by a PR.XIX operating from Malaya in April 1954, although it remained in service with the by then civilian operated Meteorological Flight until June 1957. In 1963, a PR.XIX was flown against an English Electric Lightning F.3 to develop tactics for the latter in case it had to fight Indonesian P-51 Mustangs during the period of Confrontation with that country.

The Spitfire PR.XIX was the only Griffon powered dedicated photo-reconnaissance version of the aircraft. The first was delivered in May 1944.

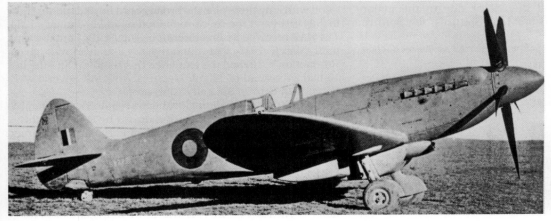

Spitfire PR.XIX SW777 was the first pressurised example. It first flew in October 1944. The aircraft was originally ordered as a prototype Spitfire Mk.21 but the contract was amended.

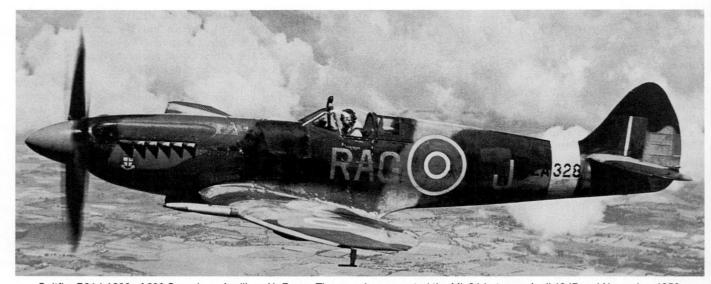

Spitfire F.21 LA328 of 600 Squadron, Auxiliary Air Force. The squadron operated the Mk.21 between April 1947 and November 1950.

Type number: 356.

First flights: Prototype – 4 October 1942 (DP851, converted from Mk.XX). Production prototype – 24 July 1943 (PP139).

Service entry: March 1945.

Powerplant: One 2,050hp (1,529kW) Rolls-Royce Griffon 61 or 65 with 2-speed/2-stage supercharger and +18lb max boost; five bladed Rotol propeller. Some with Griffon 85 and six bladed Rotol contra-props. Internal fuel 120imp gal (545 l) in forward fuselage and wing tanks; provision for 30imp gal (136 l), 45imp gal (205 l), 90imp gal (409 l) or 170imp gal (773 l) drop tank.

Dimensions: Wing span 36ft 11in (11.25m); length 32ft 8in (9.96m); wing area 244sq ft (22.6m²).

Weights: Empty 7,160lb (3,248kg); normal loaded 9,900lb (4,491kg); max overload 11,290lb (5,121kg).

Armament: Four 20mm Hispano cannon with 175rpg inboard and 150rpg outboard.

Performance: Max speed 454mph (730km/h) at 26,000ft (7,925m), 390mph (628km/h) at 4,000ft (1,220m), 407mph (655km/h) at 40,000ft (12,190m); economical cruise 284mph (457km/h); long range cruise 230mph (370km/h); max climb 4,900ft (1,493m)/min; time to 20,000ft (6,095m) 5.0min; service ceiling 43,000ft (13,105m); max range with internal fuel 580 miles (933km); max range with 90imp gal drop tank 965 miles (1,553km).

Production: 1 by Supermarine and 120 by CBAF, total 121.

Notes: At one time called the Victor due to the major differences between it and previous Spitfires, the Spitfire

PP139, first flown on 24 July 1943 and regarded as the production prototype for the Spitfire 21.

F.21 represented a new generation for the aircraft due to the incorporation of a completely new wing intended to take advantage of the power promised by the Griffon 61 engine with a two-speed/two-stage supercharger. The new fighter was considered to have enormous potential, with the result that orders covering 3,000 aircraft were placed in 1943-44.

The reality proved to be rather different, with only 120 production Spitfire 21s built after a lengthy development period which saw the aircraft suffer serious handling problems which almost resulted in it being cancelled. By the time the Spitfire 21 was sorted out and cleared for service, the European war was as good as over and the need for it had passed.

In its production form, the Spitfire 21 basically combined a Mk.VIII/XIV type fuselage (with standard canopy) with the Griffon engine, a larger diameter five bladed propeller, lengthened undercarriage struts (with the wheels fully enclosed when retracted), broad chord fin and rudder, retractable tailwheel and the new wing.

Compared to the original, the wing featured a different aerofoil section and internal structure plus wider span ailerons. The classic elliptical shape was lost but the span was only 1in (2.54cm) longer and the area 2sq ft (0.2m²) greater. A pair of 20mm cannon was mounted in each wing.

The second Spitfire IV/XX prototype was fitted with a Griffon 61 and partially modified wing (with extended tips) and reflown as the interim standard Mk.21 prototype in October 1942. The production prototype (which was more fully representative of the definitive article with the new wing – still with long tips – plus the broad chord fin and rudder) flew in July 1943 and the first production model in January 1944. The RAF received its first aircraft in August 1944 but it wasn't until March 1945 that operational clearance was given. Production F.21s had the extended tips removed and the last 15 had 24 instead of 12 volt electrical systems. Several were flown with Griffon 85s and contra-props.

The Spitfire 21's handing problems required considerable effort to resolve. The conclusion drawn by the Air Fighting Development Unit after evaluating an early aircraft seemed to signal the end of the Spitfire line by heavily criticising the aircraft's instability in the yawing plane, recommending the Mk.21 be withdrawn and replaced with Hawker Tempests or Spitfire XIVs. The AFDU concluded in December 1944 that "no further attempts should be made to perpetuate the Spitfire family". The last Spitfire F.21 was flown in December 1945.

(right) PK312, the first production Spitfire F.22. First flown in March 1945, this aircraft was originally fitted with smaller Mk.XIV/XVIII/21 tail surfaces. It received the definitive Spiteful type tail in 1946.

(below) Spitfire F.24 PK682 of 80 Squadron RAF. This was the last RAF operational squadron to fly Spitfires, relinquishing its final aircraft in Hong Kong at the beginning of 1952.

Type number: 356.

First flights: F.22 – November 1944 (SX549). F.24 – 27 February 1946 (PK313, converted from F.22).

Service entry: 1947.

Powerplant: 2,050hp (1,529kW) Rolls-Royce Griffon 61 with 2-stage/2-speed supercharger and +18lb max boost or +25lb with 150 octane fuel; five bladed Rotol propeller. Internal fuel (F.22) 120imp gal (545 l) in forward fuselage and wing tanks; (F.24) 186imp gal (845 l) in forward fuselage, rear fuselage and wing tanks; provision for 90imp gal (409 l) drop tank.

Dimensions: Wing span 36ft 11in (11.25m); length 32ft 11in (10.03m); height 13ft 6in (4.11m); wing area 244sq ft (22.6m²).

Weights: F.22 – empty 7,160lb (3,248kg); normal loaded 9,900lb (4,490kg); max loaded 11,350lb (5,148kg). F.24 – max overload 12,150lb (5,511kg).

Armament: Four 20mm Hispano cannon with 175rpg inboard and 150rpg outboard; provision for one 500lb (227kg) underfuselage and two 250lb (113kg) underwing bombs; F.24 provision for eight underwing rocket projectiles.

Performance: F.22 – max speed 450mph (724km/h) at 19,500ft (5,945m), 366mph (589km/h) at sea level or 390mph (627km/h) with +25lb boost; range cruise 230-245mph (370-394km/h); max climb 5,100ft (1,554m)/min; service ceiling 43,500ft (13,260m); max range (internal fuel) 580 miles (933km); range with drop tank 965 miles (1,553km). F.24 – max range with internal fuel 850 miles (1,368km).

Production: F.22 – 288 (Supermarine 28, CBAF 260). F.24 – 54 by Supermarine.

Notes: The Spitfire F.22 and F.24 were final developments of the line, based on the refined Mk.21 and sharing the same type number as that aircraft. The obvious external differences were the incorporation of the 'rear view' fuselage with bubble canopy and in all but the first few F.22s, the larger Spiteful type vertical and horizontal tail surfaces. Under the skin, both had the 24 volt electrical system which had been fitted to the late production Mk.21s.

The first F.22 flew in November 1944 and at one stage 627 were on order. Contract cutbacks in 1945 saw this number drastically reduced but 288 were ultimately built, mostly at the Castle Bromwich Aircraft Factory.

The first production aircraft flew in March 1945 but entry to operational service was slow, the F.22 flying with only one regular RAF squadron (No 73) from Malta in 1947-48 before re-equipping with de Havilland Vampires, while seven Royal Auxiliary Air Force squadrons operated the type between 1947 and 1951 before they also re-equipped with jets.

The Spitfire F.24 first flew in February 1946 (converted from an F.22) and production aircraft began appearing in February 1946, early aircraft converted on the production line from F.22s. The final F.24 (and the last Spitfire) was VN496, delivered in February 1948, although some more Seafires were still to come.

The Spitfire F.24 differed from the F.22 in having short barrel Hispano Mk.V cannon (instead of Mk.II) in later aircraft, electrical rather than pneumatic gun firing, additional fuel tanks in the rear fuselage and the ability to carry underwing rocket projectiles.

No 80 Squadron was the only RAF operational unit to fly the F.24, receiving its first aircraft in January 1948 in Germany and relinquishing them in December 1951 when based in Hong Kong.

Seafire I-III

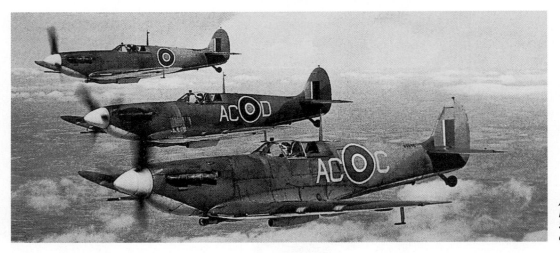

A trio of Seafire IBs with NX890 in the foreground. All Seafire IBs were conversions of Spitfire VBs.

Type numbers: IB – 340. IIC – 357. LF.IIC – 375. III – 358. Irish Mk.III conversions – 506.

First flights: IB – 6 January 1942 (AB205, converted from Spitfire VB). IIC – February 1942 (AD371, converted from Spitfire VB). III – 9 November 1942 (MA970, converted from Seafire IIC).

Service entry: IB/II – June 1942. III – April 1943.

Powerplant: IB/F.IIC – 1,470hp (1,096kW) Rolls-Royce Merlin 45 or 1,415hp (1,055kW) Merlin 46 with 1-speed/ 1-stage supercharger and +16lb max boost. F.III – 1,470hp (1,096kW) Merlin 55. L.IIC/LR.IIC/FR.III/LF.III – 1,645hp (1,226kW) Merlin 32 (+18lb max boost) or 1,585hp (1,182kW) Merlin 55M (+18lb max boost); three (Mk.I/II) or four (Mk.III) bladed Rotol propeller. Internal fuel 85imp gal (386 l) in forward fuselage tanks; provision for 30imp gal (136 l) drop tank.

Dimensions: Wing span 36ft 10in (11.23m); length 30ft 2½in (9.21m); height 11ft 5½in (3.49m); wing area 242sq ft (22.5m²).

Weights: IB – empty 5,100lb (2,313kg); loaded 6,700lb (3,039kg). L.IIC – empty 5,300lb (2,404kg); loaded 7,100lb (3,220kg). L.III – empty 6,204lb (2,814kg); loaded 7,640lb (3,465kg).

Armament: Two 20mm Hispano cannon with 60rpg (IB) or 120rpg (II/III) cannon and four 0.303in Browning machine guns in wings with 350rpg; provision for one 250lb (113kg) bomb on II, one 500lb (227kg) or two 250lb (113kg) bombs or eight 60lb (27kg) rocket projectiles on III.

Performance: IB – max speed 365mph (587km/h) at 16,000ft (4,875m); time to 20,000ft (6,095m) 7.6min; service ceiling 36,400ft (11,095m); normal range 492 miles (792km); range with drop tank 770 miles (1,239km).

Seafire IIC MB240 launching from HMS Indomitable *in 1943. The aircraft is operated by 880 Squadron Fleet Air Arm.*

L.IIC – max speed 352mph (566km/h) at 12,250ft (3,735m), 333mph (536km/h) at 5,000ft (1,525m); service ceiling 32,000ft (9,755m); normal range 493 miles (793km); range with drop tank 750 miles (1,207km).

LF.III – max speed 348mph (560km/h) at 6,000ft (1830m), 331mph (533km/h) at sea level; economical cruise 215mph (346km/h); max climb 4,160ft (990m)/min; service ceiling 32,000ft (9,755m); normal range 465 miles (748km); range with drop tank 725 miles (1,167km).

Production: IB – 166 conversions. Mk.II – 402 by Supermarine (262) and Westland (140) including 30 II/III hybrids. Mk.III – 1,220 Mk.III by Westland (870) and Cunliffe-Owen (350) comprising 100 F.III, 129 FR.III and 991 L.III.

Notes: Investigations into a carrier borne 'Sea Spitfire' for the Royal Navy Fleet Air Arm began in earnest during 1939 but the RAF's needs and political issues meant a delay, the first 'Hooked Spitfire' (a converted Mk.V) not flying until 7 January 1942. In service, what was called the Seafire provided the FAA with an excellent fighter, although the aircraft's narrow track undercarriage caused some operational difficulties. Seafires flew with 20 FAA squadrons in WW2, debuting at the Allied invasion of North Africa (Operation *Torch*) in November 1942 and subsequently operating in Sicily, Norway, the D-Day landings, Europe and the Far East against Japan.

The initial Seafire F.IBs were conversions of Spitfire VBs with standard Merlin 45 or 46 powerplant, vee-frame arrester hook and catapult spool. The 166 conversions were completed between February 1942 and July 1943 with first deliveries occurring in June 1942.

The first new build version was the Seafire II, based on the Spitfire VC with non folding wings, catapult attachment points, arrester hook, strengthened undercarriage and provision for rocket assisted takeoff gear (RATOG). Three variants were built: the F.IIC with Merlin 45 or 46; L.IIC with low rated Merlin 32 or 55M; and LR.IIC with cameras. Many had clipped wings and the undernose Vokes tropical filter. Deliveries began in June 1942.

The fully navalised Seafire III with folding wings and four bladed propeller was introduced to service in April 1943 and manufactured until mid 1945. Variants were the F.III (Merlin 55); LF.III (low rated Merlin, most with clipped wings) and FR.III with cameras. Twelve Seafire IIIs delivered to the Irish Air Corps in 1947 were refurbished and 'denavalised', in effect converted to Spitfire VC standards apart from their four bladed propellers.

The Seafire XV was the first of the navalised Spitfires powered by the Griffon engine.

Type numbers: XV – 377/386. XVII – 395.

First flights: XV – December 1943 (NS487). XVII – June 1944 (NS493, converted from Mk.XV).

Service entry: XV – May 1945. XVII – September 1945.

Powerplant: 1,815hp (1,353kW) Rolls-Royce Griffon VI with 2-speed/1-stage supercharger and +15lb max boost; four bladed Rotol propeller. Internal fuel (XV) 101imp gal (459 l) in forward fuselage and wing leading edge tanks; (XVII) additional 33imp gal (150 l) tank in rear fuselage; provision for 50imp gal (227 l) or 90imp gal (409 l) drop tank.

Dimensions: Wing span 36ft 10in (11.23m); length 32ft 3in (9.83m); wing area 242sq ft (22.5m²).

Weights: F.XV – empty 6,300lb (2,858kg); normal loaded 8,000lb (3,629kg); max loaded 9,100lb (4,128kg). XVII – empty 6,400lb (2,903kg); normal loaded 8,100lb (3,674kg); max loaded 9,600lb (4,355kg).

Armament: Two 20mm Hispano cannon with 120rpg and four 0.303in Browning machine guns with 350rpg.

Performance: XV – max speed 392mph (631km/h) at 13,500ft (4,115m); economical cruise 255mph (410km/h); max climb 4,600ft (1,402m)/min; time to 20,000ft (6,095m) 7.0min; service ceiling 37,000ft (11,280m); normal range 376 miles (605km); range with 90imp gal drop tank 903 miles (1,453km).
XVII – max speed 392mph (631km/h) at 13,500ft (4,115m), 355mph (571km/h) at 4,500ft (1,370m); normal cruise 309mph (497km/h) at 20,000ft (6,095m); max climb 4,620ft (1,408m)/min; time to 20,000ft (6,095m) 6.4min; service ceiling 37,200ft (11,340m); normal range 450 miles (724km); range with drop tank 975 miles (1,569km).

Production: XV – 390 by Supermarine (6), Westland (250) and Cunliffe-Owen (134). XVII – 232 by Westland (212) and Cunliffe-Owen (20).

Notes: The first Griffon engined Seafire, the F.XV (F.15 post war) was fundamentally a navalised version of the Spitfire XII but in detail was based on the Seafire III airframe with folding wings as well as incorporating the wing leading edge fuel tanks of later Spitfire IXs and the retractable tailwheel and pointed fin and rudder which appeared on late production Merlin-Spitfires.

The original style vee-frame arrester hook was fitted to early aircraft and the more effective 'sting' type hook on subsequent examples. This became standard on the Seafires which followed. The last 30 Seafire XVs had the 'rear view' bubble canopy and cut down rear fuselage upper decking.

Supermarine built the six Seafire XV prototypes (the first of them flying in December 1943) but production was undertaken by Westland and Cunliffe-Owen Aircraft. The Seafire XV entered Fleet Air Arm service in May 1945, just too late to see action in WWII. It served with seven FAA front line squadrons.

The closely related Seafire F.XVII (F.17) first flew in June 1944 and entered service in September 1945. Production ended in October 1946. The F.XVII shared the Griffon VI powerplant of the earlier model and featured the 'sting' hook and bubble canopy as standard plus a new and markedly superior (for carrier operations) strengthened and longer stroke main undercarriage design plus an additional fuel tank in the rear fuselage. A significant detail change was replacing the 12 volt electrical system with a 24 volt system.

The FR.17 was a 1947 conversion in which the rear fuselage fuel tank could be replaced by two F.24 cameras.

Seafire XVII SX194 of 899 Squadron Fleet Air Arm. The Mk.XVII differed mainly from the Mk.XV in having a bubble canopy and redesigned undercarriage.

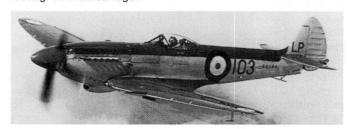

Seafire 45, 46 and 47

(left) Seafire FR.46 LA542, the second example built in late 1945. It was used for various tests and trials.
(below) TM379, the proto- type Seafire 45, first flown in August 1944.

Type number: 388.

First flights: 45 – August 1944 (TM379). 46 – 8 September 1944 (TM383). 47 – 25 April 1946 (PS944).

Service entry: 45 – November 1946. 46 – July 1946. 47 – mid 1947.

Powerplant: 45/46 – 2,050hp (1,529kW) Rolls-Royce Griffon 61 with 2-speed/2-stage supercharger and +18lb max boost; five bladed Rotol propeller. 47 – 2,145hp (1,599kW) Griffon 87 or 2,350hp (1,752kW) Griffon 88 (+18lb boost or +25lb with 150 octane fuel); six bladed Rotol contra-props. Internal fuel (45) 120imp gal (545 l) in forward fuselage and wing tanks; (46/47) 152imp gal (691 l) in forward fuselage, rear fuselage and wing tanks; provision for drop tanks.

Dimensions: 45/46/47 – Wing span 36ft 11in (11.25m); wing area 244sq ft (22.6m²). 45 – length 32ft 8in (9.96m). 46/47 – length 34ft 4in (10.46m); height 12ft 9in (3.88m).

Weights: 45 – empty 7,100lb (3,220kg); max loaded 10,000lb (4,536kg). 46 – empty 7,500lb (3,402kg); normal loaded 9,900lb (4,491kg); max loaded 11,400lb (5,171kg). 47 – empty 7,625lb (3,459kg); normal loaded 10,700lb (4,853kg); max loaded 12,750lb (5,783kg).

Armament: Four 20mm Hispano cannon with 175rpg inboard and 150rpg outboard; provision for 250lb (113kg) bombs or depth charges, eight rocket projectiles on Mk.46/47.

Performance: 45 – max speed 446mph (718km/h) at 26,000ft (7,925m); maximum climb 5,300ft (1,615m)/min; service ceiling 43,000ft (13,105m); range 580 miles (933km). 47 – max speed 452mph (727km/h) at 20,500ft (6,250m), 353mph (568km/h) at sea level, 405mph (652km/h) at 9,500ft (2,895m); range cruise 215-235mph (346-378km/h); initial climb 4,800ft (1,463m)/min; time to 20,000ft (6,095m) 4.9min; service ceiling 43,100ft (13,135m); max range with 90imp gal drop tank 940 miles (1,513km).

Production: Mk.45 – 50 by CBAF. Mk.46 – 24 by Supermarine. Mk.47 – 90 by Supermarine.

Notes: The three final Seafire variants equated to navalised versions of the last three Spitfire models – the Seafire 45 to the Spitfire 21, Seafire 46 to Spitfire 22 and Seafire 47 to Spitfire 24, except in the case of the latter the fitting of a more powerful Griffon 87 or 88 engine driving counter-rotating propellers. A small number of Mk.45/46s were tested with this engine/propeller configuration.

All three utilised the new wing which had first been applied to the Spitfire 21 and its four cannon armament, the Seafire 45 and 46 retaining the non folding wing of their land based counterparts which made them unsuitable for use from several aircraft carriers. The Seafire 47 introduced single fold wings, manually operated in early aircraft and then hydraulically actuated.

Externally, the Seafire 45 retained the standard canopy and broad chord fin and rudder of the Spitfire 21 while the Mks.46 and 47 had bubble canopies and the enlarged Spiteful type vertical and horizontal tail surfaces of the Spitfire 22/24. 'Sting' hooks were fitted and the Mk.47 introduced a 24 rather than 12 volt electrical system. All three were built in standard fighter 'F' and fighter-reconnaissance 'FR' versions with rear fuselage camera. Most Mk.47s were completed to FR.47 standards.

Only the F/FR.47 entered front line Fleet Air Arm squadron service and then with just two units. No 800 Squadron's Mk.47s were the only of its type to see action. Based aboard HMS *Triumph* but operating from land in Singapore, they conducted rocket attacks against communist terrorists in late 1949 before spending a three month tour of duty (from July 1950) flying ground attack sorties from *Triumph* during the Korean War.

The final Seafire FR.47 (VR972) was also the last of the Spitfire/Seafire line, delivered in January 1949.

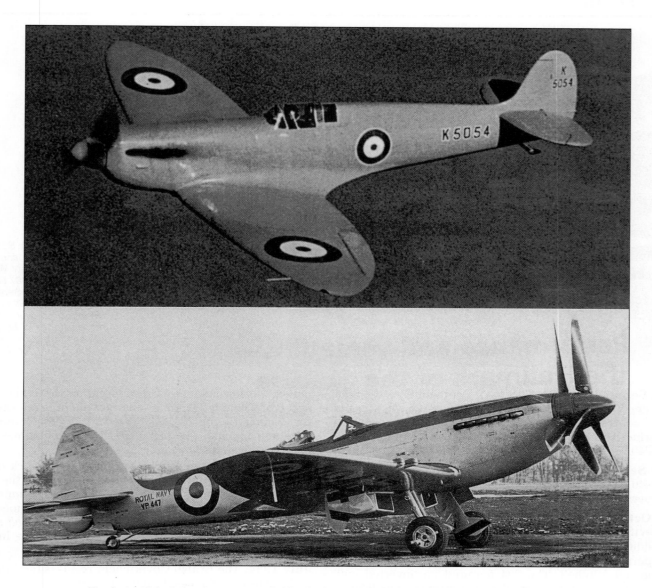

The first and the last – the prototype Spitfire (top) of 1936 and the Seafire FR.47 of a decade later.

SPITFIRE CHRONOLOGY

Supermarine S.6B S1596, the second of two and the one which set a new World Speed Record of 407.5mph (655.8km/h) on 29 September 1931, flown by Flt Lt G H Stainforth. Just over a fortnight earlier, Flt Lt J N Bootham had permanently won the Schneider Trophy for Britain in S1595.

13 September 1931: Flt Lt J N Bootham wins the Schneider Trophy permanently for Britain by winning the 12th contest unopposed at Spithead. Flying the Supermarine S.6B S1595, Bootham records an average speed of 340.08mph (547.29km/h) over the course.

29 September 1931: Flt Lt G H Stainforth sets a new World Speed Record of 407.5mph (655.8km/h) in the Supermarine S.6B S1596.

1 October 1931: The British Air Ministry circulates Specification F.7/30 for a new fighter, this attracting the building of prototypes by various companies including Supermarine's Type 224, the first fighter designed by Reginald Mitchell. The 224 featured an 'inverted gull' wing, fixed and 'trousered' undercarriage, four 0.303in machine guns (one in each undercarriage fairing and one on each side of the fuselage above the wing fairings), an open cockpit and a Rolls-Royce Goshawk engine with evaporative water cooling. This complicated system required a series of condensers in the wings to which steam was carried from the engine cooling jacket, onto condensation collecting tanks and then back to the cooling jacket.

The specification called for a maximum speed of 195mph (315km/h), a landing speed of 60mph (96km/h) and a time to height performance of 8½ minutes to 15,000ft (4,570m). Designs from Supermarine, Blackburn, Gloster, Westland, Parnall, Boulton Paul and Hawker were submitted to the Air Ministry in February 1932 and of those the first three were awarded a contract to build a prototype.

April 1933: First detailed drawings of Rolls-Royce P.V.12 (later Merlin) vee-12 engine issued.

15 October 1933: First Rolls-Royce P.V.12 engine bench tested.

February 1934: First flight of the Supermarine Type 224 (K2890). Despite easily beating the specification requirements, performance was disappointing, both the top speed of 238mph (383km/h) and rate of climb inferior to the winner of the F.7/30 contract, the Gloster SS 37 Gladiator biplane. The name 'Spitfire' is sometimes applied to this aircraft.

May 1934: Reginald Mitchell begins design work on a new fighter incorporating some of the features of the 224 but with evolutionary modifications intended to improve its performance. This work was originally performed as a private venture by Supermarine and was the first step in the process that led to the Type 300 Spitfire in its recognised form.

July 1934: The British Government releases the first of its Expansion Schemes designed to cope with the rise of Hitler in Germany and his avowed intention to re-arm and re-equip the *Luftwaffe*. Expansion Scheme A called for a Royal Air Force strength of 28 fighter and 43 bomber squadrons.

July 1934: Supermarine proposes the Type 300 fighter to the Air Ministry, based on the 224 with Goshawk engine but with enclosed cockpit, retractable undercarriage and a straight tapered and 'unbent' wing of reduced span and area. Maximum speed is estimated at 270mph (434km/h) but Air Ministry reaction is lukewarm.

July 1934: P.V.12 engine completes 100 hour Type Test. At this stage the engine is rated at 625hp (466kW) for takeoff and 790hp (589kW) at 12,000ft (3,660m) with +2lb boost.

19 July 1934: The British Air Ministry holds a conference on the issue of aircraft armament attended by a number of senior officers. Among them was Sqn Ldr R S (later AM Sir Ralph) Sorley of the Air Ministry's Flying Operations 1 unit, responsible for various Air Staff requirements. Sorley pressed the case for the eight gun fighter, arguing that a lesser number would be incapable of destroying or badly damaging an enemy aircraft. With support from others present, Sorley was able to

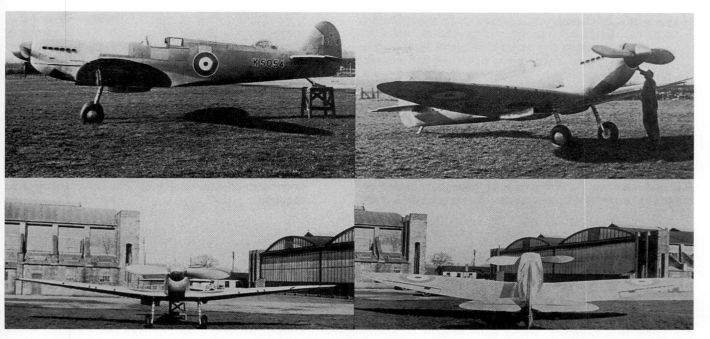

A walk around the Supermarine Type 300 prototype (K5054) in March 1936, shortly before the aircraft's maiden flight. Note the aircraft has yet to have its main undercarriage fairings fitted and is unpainted.

convince the conference that this was the way to go, with the result that both the Spitfire and Hawker Hurricane first appeared armed with eight 0.303in Browning machine guns.

November 1934: Supermarine offers a refined Type 300 with a hint of the definitive Spitfire's lines evident. The straight tapered wing is retained but the four guns are now buried in the wings. Alternatives to the Goshawk engine are sought, Supermarine concluding an agreement with Rolls-Royce to supply one of its new P.V.12 liquid cooled vee-12 engines for the prototype when it is built. The P.V.12 would later be named 'Merlin'.

6 November 1934: Reginald Mitchell is authorised by the Vickers board of directors to proceed with detail design work on the Type 300 with a view to building a company funded prototype. The approval resulted largely from the enthusiasm for the project demonstrated by Sir Robert McLean, the managing director of Vickers (Aviation) Ltd.

November 1934: It was during the third week of this month that Reginald Mitchell decided to use an elliptical wing on the Type 300 rather than the straight taper design previously utilised, the idea being to provide sufficient depth for the retractable undercarriage and the guns and their associated ammunition boxes. The earliest elliptical wing drawings show a perfect ellipse design, this evolving into the characteristic Spitfire 'distorted ellipse' shape with less curved leading edge. This first appeared in a December 1934 drawing.

The final wing planform was dictated by the need to install eight machine guns rather than aerodynamic advantages, although the low thickness/chord ratio of the final product proved to be remarkably efficient. Right from the start it was recognised that the elliptical wing would be more difficult to manufacture than a normal

tapered design, and so it proved to be. Supermarine's chief draughtsman, Joseph Smith (later chief designer) was responsible for the wing's structural design.

1 December 1934: The Air Ministry authorises manufacture of the prototype Supermarine Type 300 fighter, the cost estimated at £10,000. Specification F.37/34 was raised to cover the contract, which stipulated the aircraft to be ready by October 1935. Vickers' approval of the project the previous month had been as a private venture; now it was a government venture with a relatively small amount of company expenditure.

28 December 1934: Air Ministry Specification F.37/34 written around the Supermarine Type 300 for "an experimental high-speed single-seat fighter substantially as described in Supermarine Specification No 425a...."

April 1935: The Rolls-Royce P.V.12 engine first flown in a Hawker Hart testbed.

April 1935: Specification F.10/35 issued, calling for "not less than six guns, but eight guns are desirable...". This was received just as the Type 300 mockup (with only four guns) was completed. Mitchell quickly designed an eight gun installation and the Air Ministry contract for the prototype was amended. Interestingly, F.10/35 also called for a *reduction* in fuel capacity, from 94imp gal (427 litres) to 66imp gal (300 litres), although the amended Type 300 contract called a for a minimum capacity of 75imp gal (341 litres).

26 June 1935: The name 'Spitfire' appears for the first time on Supermarine's reports on the Type 300, although at this stage the Air Ministry had no knowledge of it. The name had been thought up by the chairman of Vickers (Aviation) Ltd, Sir Robert McLean. He wanted a name starting with an 's' to establish the desired alliteration and sibilance and also required something with a bit of venom in it. The alternatives 'Shrew' and 'Shrike' were

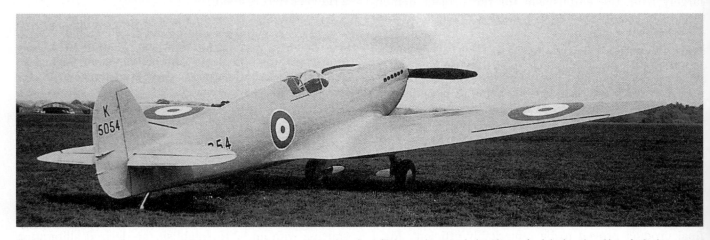

Three more views of the prototype, this time a few months after first flight and around the time of trials by the Aircraft & Armament Experimental Establishment. Note the exhaust system and the 'flap' covers for mainwheels, these covering the wheels completely when retracted. Note also the smaller rudder horn balance.

SUPERMARINE HERITAGE

The name Noel Pemberton-Billing is not one readily associated with the great pioneers of British aviation – legends such as Geoffrey de Havilland, Alliot Verdon-Roe, Harry Hawker, Robert Blackburn, the Short brothers or Thomas Sopwith – but it is significant nevertheless because it is the name behind what became The Supermarine Aviation Works Ltd.

Pemberton-Billing was a wealthy, adventurous and outspoken type who had made a name for himself as a yachtsman and in the new field of motoring. His personality is well summed up by the circumstances surrounding his learning to fly in 1913. He wagered £500 that he could earn his aviator's licence not just in a single day, but before breakfast! In order to achieve his aim and win the bet, Pemberton-Billing bought a Farman biplane and hired an instructor from the Vickers flying school at Brooklands.

Having mastered the art of flying, Pemberton-Billing's next aviation quest was to build aeroplanes. Using his boat building factory at Woolston (Southampton) as the basis, his stated aim was "to build boats that could fly, rather than aeroplanes that could float."

The firm of Pemberton-Billing Ltd was registered in 1914 and the company's first product – the PB.1 biplane flying boat – appeared shortly afterwards. The PB.1 was promoted as being a 'super-marine' craft. The name 'Supermarine' was also used as the factory's telegraphic address but unfortunately, the PB.1 was not so super – it never flew.

A couple of other designs did succeed in flying but in 1916 – after making severe criticism of the conduct of WWI – Pemberton-Billing decided to stand for Parliament to give his opinions some weight and also joined the Royal Naval Air Service.

Shortly afterwards, the company was restructured and started performing subcontract work for other manufacturers. Some new faces joined the firm, among them a young engineer called Reginald Mitchell.

Pemberton-Billing then sold his interest in the company which changed its name to The Supermarine Aviation Works Ltd. And the rest – as they say – is history.

One of Pemberton-Billing Aircraft's early efforts, the remarkable PB.31E Night Hawk, intended to counter the Zeppelins which were raiding Britain by night in World War I.

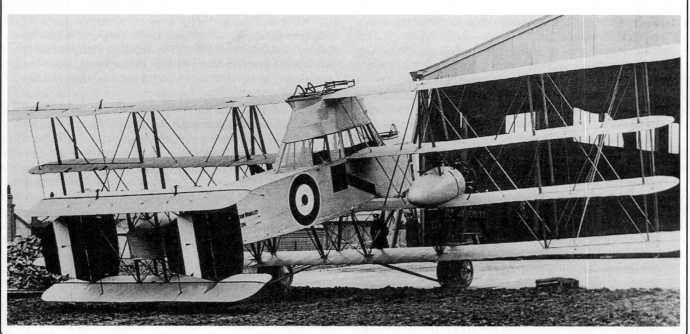

also considered for a time but it was not until after the prototype flew in March 1936 that the Air Ministry somewhat reluctantly agreed to adopt Sir Robert's first – and inspired – choice. Reginald Mitchell was also apparently less than enamoured with it....

August 1935: The first metal is cut and the Type 300 design is more or less settled to the configuration and mechanical specification that would appear with the prototype early in 1936. There were some detail changes such as increasing the fuel capacity from 75imp gal

(341 l) to 85imp gal (386 l), the wing span was reduced slightly, the fuselage aft of the widened cockpit was changed to an oval section and a small glazed section aft of the canopy incorporated.

November 1935: By now, construction of the prototype was well advanced with the fuselage nearly complete, engine installed and wings skinned. Some design matters still had to be resolved, notably the position of the oil cooler, which at this stage was under the engines. It was soon relocated to under the port wing but the final positioning

The one and only Supermarine Type 224 fighter (K2890), Supermarine's unsuccessful contender for the F.7/30 contract.

of the radiator – under the starboard wing – had been decided. Previously, this was under the port wing.

It was estimated that the prototype would fly for the first time in early January 1936 but the work associated with relocating the oil cooler was largely responsible for a two months delay.

18 February 1936: The Spitfire prototype is completed and initial engine ground running is conducted at the Woolston plant. On the completion of these tests, the aircraft was dismantled and trucked to Eastleigh aerodrome for reassembly in preparation for its first flight.

5 March 1936: First flight of the unpainted prototype Supermarine Type 300 (K5054), conducted by Capt J 'Mutt' Summers, Vickers' chief test pilot, from Eastleigh aerodrome near Supermarine's main factory at Woolston (Southampton) with power provided by a 990hp (738kW) Rolls-Royce Merlin C driving a two bladed fixed-pitch wooden propeller. The 15 minute flight was conducted with the undercarriage down and a fine pitch propeller fitted.

The takeoff was performed at an angle of 35 degrees across the wind in anticipation of a strong tendency to swing due to the effects of torque – it was easily controlled. After the flight, Summers uttered the words "I don't want anything touched", these being subsequently misinterpreted to mean the aircraft was perfect in every way and needed no modification – all after a 15 minute assessment! What he meant was that there were no unserviceabilities that needed to be dealt with before the next flight.

25 March 1936: Supermarine test pilot George Pickering flies the prototype for the first time.

26 March 1936: Flt Lt Jeffrey Quill – the assistant to Vickers' chief test pilot 'Mutt' Summers – flies the prototype Spitfire for the first time, thereafter assuming most of the responsibility for development and flight testing of the aircraft until near the end of WWII. Early testing revealed disappointment with the aircraft's top speed (about 335mph/539km/h at optimum height) but trying different propellers and introducing other 'tweaks' increased this by more than 10mph (16km/h) by the time K5054 was released for official testing.

10 April 1936: Following initial flight trials, the Spitfire prototype goes back to the workshop for modifications. These included reducing the size of the rudder horn balance, lowering the carburettor intake to increase ram air pressure and painting the aircraft in the blue-grey colour scheme which has caused identification problems for enthusiasts and historians ever since.

5 May 1936: The prototype Spitfire's disappointingly low maximum speed was thought to be as a result of the original propeller's tips reaching a high Mach number and therefore having compressibility problems. A new propeller with modified tips was fitted and the maximum speed increased to 348mph (560kmh), just short of the target 350mph (563km/h) but acceptable.

3 June 1936: An initial order for 310 Type 300 Spitfire Is placed, in accordance with the Air Ministry's Expansion Scheme F which required the delivery of 300 Spitfires and 600 Hawker Hurricanes within three years. Specifically, the Spitfire contract required the first delivery on 12 October 1937 and the last in March 1939 – these deadlines were not met, the programme running many months behind schedule in 1938 and 1939.

A new specification (16/36) was drawn up to cover the production standard Spitfire I and the 33 items of differences between it and the prototype. These included provision for a three bladed metal propeller, flare dropping capability, increased fuel capacity, landing lights, a tailwheel (instead of skid), detachable wingtips and increasing the height of the cockpit canopy for additional headroom. Expansion Scheme F called for an RAF strength of 1,736 front line aircraft by 1939 including 30 home based fighter squadrons. Substantial reserves were also allowed for.

10 June 1936: The name 'Spitfire' is officially approved for the Supermarine Type 300 fighter.

18 June 1936: The Spitfire makes its first 'semi public' appearance before an invited audience of 300 at Supermarine's Eastleigh facility. Most were from the aviation industry but their chance to see the aircraft in the air was curtailed when an oil line broke, forcing Jeffrey Quill to perform a skilful emergency landing.

27 June 1936: The prototype Spitfire makes its first appearance for the general public at the RAF Pageant, Hendon. The Hawker Hurricane prototype also appeared at the pageant in the New Types Park. The two new fighters flew together, the Spitfire piloted by Flt Lt J H Edwardes-Jones.

30 June 1936: 'Mutt' Summers flies the prototype at the Society of British Aircraft Constructors (SBAC) display at Hatfield.

July 1936: Testing of K5054 at the Aircraft & Armament Experimental Establishment (A&AEE) at Martlesham Heath revealed a maximum speed of 349mph (562km/h) at 16,800ft (5,120m) and a time to height figure of 5min 42sec to 15,000ft (4,570m). A&AEE pilots praised the aircraft's handling qualities: "The aeroplane is simple and easy to fly and has no vices.... the controls are well harmonised and appear to give an excellent compromise between manoeuvrability and steadiness for shooting.... in general the handling of this aeroplane is such that it can be flown without risk by the average fully trained service fighter pilot."

11 July 1936: During testing at Martlesham Heath, Hugh Edwardes-Jones takes the prototype up to its so far greatest altitude – 37,400ft (11,400m) – a height reached in 37 minutes.

14 July 1936: Formation of Royal Air Force Fighter Command, its early organisation comprising two Fighter Groups (Nos 11 and 12), No 22 Army Co-operation Group and the civilian manned Observer Corps. Fighter strength comprised Hawker Fury and Bristol Bulldog biplanes which were in the process of being replaced by the Gloster Gauntlet and Hawker Demon – also biplanes. The first monoplane Hawker Hurricane squadron was No 111 at Northolt, re-equipped from December 1937. Commanded by Air Marshal Sir Hugh Dowding KCB CMG, Fighter Command's structures and philosophies were put into place under his strong leadership – including his quick recognition of the benefits of radar – with the result that Britain was able to defend itself against the *Luftwaffe* in 1940.

18 July 1936: The prototype is test flown with the hinged undercarriage flap/doors removed. These meant that the wheels were completely enclosed when retracted but they were causing problems by often refusing to close properly. Their removal made no difference to the aircraft's performance and were therefore discarded.

25 July 1936: First production Merlin I delivered. Rated at 1,030hp (768kW), its only application was the first 136 Fairey Battles.

22 September 1936: Acceptance tests on the 1,030hp (768kW) Merlin G (or II) completed. This was the Merlin variant fitted to early Spitfire Is and Hurricane Is.

3 December 1936: The prototype reflies after spending nearly three months in the workshop undergoing modifications including the fitting of armament.

20 December 1936: The prototype flies for the first time fitted with a de Havilland variable pitch propeller.

22 December 1936: The prototype is fitted with radio.

8 January 1937: The prototype Spitfire flies for the first time with a tail wheel instead of the original skid. A then secret decision by the Air Ministry to start building sealed all weather runways at RAF bases was behind the change, the skid having been intended for the grass airfields then in use. Two types of tailwheel were tested, one with a single wheel (as subsequently adopted) and the other with two smaller wheels.

January 1937: By the end of the month the prototype had recorded 95hrs 50min flying time and had completed a large number of manufacturer's tests and trials. Among them: spinning; modified elevator gearing; effect of gun installation on handling; installation of 1,030hp (768kW) Merlin F; various oil coolers; two and four bladed propellers, fixed and variable pitch; tailwheels (instead of the original skid); modified air intakes and exhaust manifolds; and revised aerial masts.

January 1937: The RAF's Expansion Scheme F – under which the initial Spitfire I orders had been placed – is deemed to be inadequate given the *Luftwaffe's* rapid growth and Scheme H is proposed instead. This called for increasing the RAF's home based front line force to about 2,500 aircraft after April 1939 (mainly by reducing the number of reserves) but it is not accepted as the threat from Germany increased. Schemes J

(above) The prototype was painted in operational brown/green camouflage in September 1937. By then, it had been modified to a standard close to that of early production Spitfires including the fitting of armament. (below) The Spitfire prototype's first accident on 22 March 1937 when a 'blown' engine forced Flt Lt J F McKenna to perform a wheels up, dead stick landing in a field. The aircraft was out of action for nearly six months.

and K were drawn up and rejected during 1937 with Scheme L finally being settled on in March 1938 when German troops entered Austria. This planned a rapid expansion of the RAF's strength, taking it to 12,000 aircraft in two years should war break out.

23 February 1937: Full handling trials of the prototype conducted at the A&AEE, Martlesham Heath.

March 1937: Gun firing trials conducted on the prototype by the A&AEE reveal no problems with the pneumatically operated system on the ground and at low altitudes, but at high altitudes the breech oil was freezing and jamming the guns. The problem was exacerbated by the fact that the guns had to be cocked prior to flight, leaving the breeches open. As the aircraft lost altitude the oil thawed out and on one occasion three of the guns fired on landing as a result of the touchdown jolt.

22 March 1937: The prototype's first accident when the engine suffered oil starvation during a high 'g' turn, forcing Flt Lt J F McKenna to switch it off and conduct a wheels up landing in a field. Damage to the airframe was minor but the engine was found to have 'blown', with failed connecting rods and big end bearings punching four holes in the crankcase. When it comes to matters of aviation, the popular press was obviously just as inept in 1937 as it is today.... "The Supermarine Spitfire prototype was deliberately crashed to ascertain the strength of the airframe", according to one caption writer!

11 June 1937: The death, in Vienna, of the Spitfire's designer, Reginald Joseph Mitchell CBE AMICE FRAeS, aged 42 after a long battle with cancer. He never saw his most famous creation enter service but at least died in the knowledge that it had been ordered in substantial quantities for the RAF.

10 August 1937: First production Merlin II for early Spitfire Is and Hurricane Is delivered.

9 September 1937: The prototype is reflown after its March accident with further modifications incorporated. Externally, it looked like one of the production Spitfire

A general view of the prototype's cockpit with the large undercarriage hand pump lever obvious on the right hand side.

Is which would follow as it was repainted in the dark green/brown earth camouflage scheme that would be applied to those aircraft, but inside it was still a long way from production standards.

19 September 1937: The quest for more speed resulted in the flight testing of a new ejector exhaust pipe system.

The prototype photographed at Eastleigh in late 1937 after repairs, modification and repaint. Note the exhaust system, tailwheel, guns, radio mast and slightly revised upper cowling contours compared to its earlier form. The small bulge just forward of the fin is the spin recovery parachute shackle.

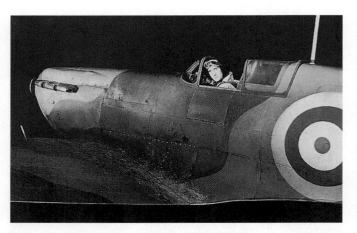

Jeffrey Quill in the cockpit of the prototype during night flying trials in early 1938. Excessive glare from the exhaust pipes was a problem.

THE DESIGNERS – MITCHELL and SMITH

Reginald Joseph Mitchell joined Supermarine at the age of 21 in 1916 as the assistant to the company's principal, Hubert Scott-Paine. As a young and obviously talented engineer, he rose quickly within the company and was given considerable responsibility at an early age.

He was appointed chief designer in 1919 and chief engineer the following year, in addition becoming a director in 1927. Mitchell was responsible for the design of most of Supermarine's aircraft during this period and into the 1930s, his skills covering such disparate types as the various flying boats produced by the company to the high speed S.4, S.5, S.6 and S.6B floatplanes which won the Schneider Trophy for Britain and set new world speed records.

Some find it difficult to believe that designs such as the contemporary Walrus biplane amphibian and the Spitfire could come from the same pen, but although vastly different in just about every respect of their design and performance, they had the common theme of being very effective in the roles for which they were designed.

In 1933, Reginald Mitchell contracted the cancer which finally killed him. Various treatments over the next four years failed and he finally passed away at the tragically early age of 42 on 11 June 1937.

Mitchell's position of chief designer for Supermarine was taken over by his assistant, Joseph Smith. After working for the Austin Motor Company, Smith joined Supermarine in 1921 at age 23 and was appointed chief draughtsman in 1926. Intimately involved with the Type 300 Spitfire programme from the start (he took Mitchell's designs and translated them into finished drawings), Smith was well placed to take over when Mitchell passed away.

His contribution over the next decade and especially during the war years was extraordinary as he was responsible firstly for transforming the Spitfire from a prototype into a production machine and then development of the myriad of marks and models which followed, all under the immense pressures resulting from war.

With careful setting of the rearwards facing system, this was found to produce some residual thrust at no cost and increased the maximum speed of the prototype by another 10mph (16km/h) to 360mph (579km/h) at optimum altitude.

23 October 1937: The prototype Spitfire is officially taken on RAF charge. Much of the following few months is taken up with more trials, mostly covering armament, gun camera and night flying. Additional operational equipment is installed including a reflector gunsight, camera gun and gun heating ducts.

January 1938: Night flying tests with the prototype begin, the major problem being excessive glare from the exhaust pipes.

March 1938: Two months before the maiden flight of the first production Spitfire I, the manufacturing programme was falling behind schedule with 35 fuselages completed by Supermarine's Woolston works and 25 of these transferred to Eastleigh for final assembly. Wings were a problem – only four sets had been delivered by General Aircraft Ltd, although the situation improved when Supermarine began making its own Spitfire wings later in 1938.

15 March 1938: The prototype suffers its second mishap during night flying trials intended to determine the amount of glare being generated by the exhaust system. The pilot, Sgt Wareham, performed two missed approaches and on the third attempt to land overshot the runway, putting the Spitfire into soft ground where it tipped onto its nose. The propeller was broken and the cowling bent, but the aircraft was quickly flying again.

24 March 1938: Just nine days after the incident noted above, the prototype came to grief once more, again during night flying trials. This time the aircraft suffered a heavy touchdown and bounced a considerable distance, at the end of which the port wing touched the ground. A ground loop followed, forcing the port undercarriage leg

The first official photograph of operational Spitfire Is, taken on 31 October 1938. The aircraft are from the first unit equipped with the Spitfire, No 19 Squadron based at Duxford in Cambridgeshire. The numbers were painted on the tails especially for the occasion and the front aircraft (K9794) is flown by the squadron's commanding officer, Sqn Ldr Henry Cozens.

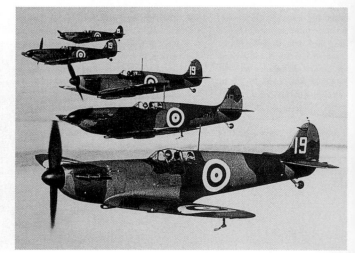

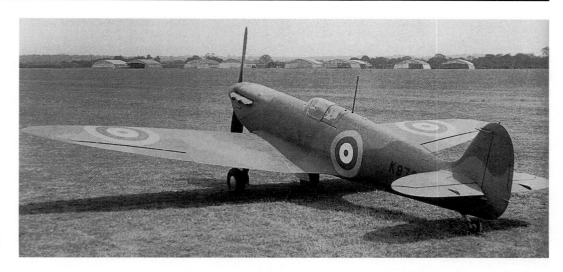

K9787, the first production Spitfire Mk.I. It first flew on 14 May 1938.

No 66 Squadron at Duxford was the second RAF unit to put the Spitfire into service, receiving its first aircraft on 28 October 1938.

An early production Spitfire I carrying 19 Squadron's pre war 'WZ' code. (via Neil Mackenzie)

With its highly boosted engine, the High Speed Spitfire was a finicky machine. It had to be started on one set of (cold) plugs and when warm, shut down so a set of hot plugs could be installed.

up through the wing. The prototype was out of action for nearly seven months while being repaired, not flying again until October.

May 1938: Early investigations begin into the lengthy process of adapting the Spitfire for carrier based naval operations, the Admiralty approaching Fairey about the possibility of building the aircraft.

May 1938: Car manufacturing magnate Lord Nuffield (William Morris) approached by the Air Ministry on the subject of bringing his mass production expertise into aircraft manufacture. The result was the Castle Bromwich Aircraft Factory in Birmingham, the major source of Spitfire production from 1940 (see sidebar story).

14 May 1938: Maiden flight of the first production Spitfire I (K9787) at Eastleigh, nearly a year later than originally planned. After undergoing type trials at the A&AEE in July, the aircraft went to Farnborough for radio trails and was subsequently converted to photo-reconnais-

sance configuration for service with the PRU. It was lost in June 1941.

1 July 1938: First production Rolls-Royce Merlin III (for later Spitfire Is and Hurricane Is with three bladed constant-speed propellers) delivered.

13 July 1938: Air Ministry Certification of Design for the Spitfire awarded. This meant that deliveries to the RAF could begin.

30 July 1938: The sixth production Spitfire I (K9792) becomes the first Spitfire to enter RAF service when it is delivered to Central Flying Establishment (CFE) at Cranwell for evaluation by instructors.

4 August 1938: No 19 Squadron at Duxford becomes the first RAF operational unit to receive the Spitfire, the third production Mk.I K9789. K9790 joins the squadron three days later. The initial delivery rate was slow, averaging one per week.

20 September 1938: The first Spitfire writeoff: Plt Off G Sinclair and K9792 come to grief during landing at Duxford due to a stub axle breaking after a heavy landing. The aircraft flipped onto its back but the pilot was unhurt. It was his first flight in a Spitfire!

October 1938: Reorganisation of the Vickers and Supermarine companies occurs, with Vickers (Aviation) Ltd and the Supermarine Aviation Works (Vickers) Ltd subsumed into Vickers-Armstrong Ltd. Vickers (Aviation) Ltd had purchased Supermarine in 1928 and in 1931 the company name had been changed to Supermarine Aviation Works (Vickers) Ltd. From 1938, Vickers-Supermarine was the common name of the company.

12 October 1938: Following repairs resulting from the accident of 24 March (see above), the prototype resumes flying, conducting gun heating trials at Martlesham Heath. The tests involved firing the guns at heights of up to 30,000ft (9,145m). By now, 13 production Spitfire Is

VALUE FOR MONEY

Cost overruns in aircraft programmes are not a modern phenomena – the initial contract for 310 Spitfire Is estimated the cost of each airframe at £4,500 bare (without engine, guns, radios, instruments etc). Their actual cost was £6,033 each, a 34 per cent increase. Not that too many British taxpayers were complaining during the summer of 1940!

Nor would they have been complaining too much about the cost of the prototype, which up to the time of its demise in September 1939 owed the taxpayers of the nation £12,478 – has there ever been a better government funded investment?

The High Speed Spitfire, first flown on 10 November 1938.

had been flown and a total of 49 will have been delivered to the RAF by the end of the year – well behind schedule.

28 October 1938: No 66 Squadron receives its first Spitfire (K9802), joining its fellow Duxford based unit (19 Squadron) in operating the new fighter.

10 November 1938: First flight of the Type 323 High Speed Spitfire, built for an attempt on the World's Landplane Speed Record. The aircraft was converted on the production line from Spitfire I K9834 and given the Class B registration N.17 (see *High Speed Spitfire* breakout box).

17 November 1938: Jeffrey Quill flies the 28th production Spitfire I (K9814) from London (Croydon) to Paris (Le Bourget) in an unofficial record time of 42.5 minutes, setting off considerable aviation press speculation of the aircraft's speed capabilities. On arrival, the Spitfire was dismantled and trucked to the Grand Palais for an indoor static display at the *Salon de l'Aeronautique International* which was held every year. The return flight was even faster, taking only 41 minutes.

December 1938: Supermarine receives its second order for Spitfire Is, covering 200 aircraft and resulting from the alarm bells which sounded following the Munich Crisis.

December 1938: By the end of the year, 49 Spitfire Is had been delivered to the RAF.

PROTOTYPE DEVELOPMENT

By modern standards, it seems incredible that the prototype (K5054) was the only Spitfire flying for over two years before the first production aircraft took to the air. As a result, this aircraft underwent considerable development between its first flight in March 1936 and its demise in a landing accident in September 1939.

As first flown it was powered by a 990hp (738kW) Merlin C driving a two bladed, fixed pitch and wooden de Havilland propeller. Other features included a tail skid, flush exhaust ports and a flat cockpit canopy.

Testing, evaluation and development led to increasing the flap deflection angle to 90deg to reduce the tendency to float on landing (Spitfires only had two flap positions – fully up or fully down), incorporation of a small knockout panel in the side of the canopy to equalise air pressure and allow the canopy to be more easily jettisoned at high speeds, the area of the rudder aerodynamic balance was reduced, ejector exhaust ports were fitted to give an extra 10mph (16km/h) speed and the original full covers for the main wheels (when retracted) had their lower hinged portions removed.

The eight gun armament was fitted in December 1936 and officially tested in early 1937, jamming at high altitudes proving to be a problem when the breech oil froze. A satisfactory gun heating system took a considerable amount of time to develop, the problem not being finally solved until late 1938.

Powerplant development resulted in the fitting of a 1,030hp (768kW) Merlin G in 1937 – equivalent to the production Merlin II – while the tail skid was replaced with a wheel in early 1937 and the production style radio mast was fitted just behind the cockpit. Following repairs resulting from an accident, the prototype emerged in September 1937 with more modifications and the green/brown camouflage scheme that would soon adorn production aircraft.

Subsequent testing covered the trial of different propellers, exhaust systems (to reduce glare during night flying), the troublesome gun heating system and numerous items of operational equipment.

1939

January: Rolls-Royce begins development of a 37 litre vee-12 engine (eventually the Griffon) based on the 'R' engine used in the Supermarine S.6/S.6B Schneider Trophy and World Speed Record aircraft of 1929-31.

20 February: Supermarine issues a specification for an 'Improved Spitfire' with greater performance and upgraded armament including two cannon and four machine guns or four cannon. Notable changes for what would develop into the Type 348 Spitfire III prototypes included a more powerful RM.2SM (Merlin X) engine, reduced span wings, additional fuel capacity and retractable tailwheel. A maximum speed of 400mph (644km/h) at 21,000ft (6,400m) was predicted.

March: Spitfire I (L1007) begins flight trials with 20mm Hispano cannon mounted in the wings as the prototype Mk.IB. Testing with the A&AEE at Martlesham Heath and the Air Fighting Development Unit (AFDU) at Northolt revealed the guns to be unreliable, quickly jamming at low temperatures and necessitating the installation of heating. Wing flexing during high 'g' manoeuvres also contributed to the problem.

April: Delivery of the 150th Spitfire I – six months behind schedule.

April: The Portuguese Government orders 15 Spitfire Is for the *Arma de Aeronautica* of which the sale of 12 is approved by the British Foreign Office. The outbreak of war results in their cancellation, Portugal finally receiving 18 second hand Mk.Is in November 1942 and 33 ex RAF Mk.VBs from October 1943.

May: Rolls-Royce opens a second aero engine manufacturing plant at Crewe, followed by a third near Glasgow in October 1940. The Ford Motor Company's Manchester factory also began manufacturing Merlins in 1941.

19 May: Delivery of the first Spitfire I with Merlin III engine (K9980, the 194th aircraft) to 611 Squadron at Speke (Liverpool).

The Spitfire assembly line at Woolston in early 1939 with five Mk.Is visible and a stock of Merlin engines ready in the foreground. Note the upper fuel tank in front of the nearest aircraft, ready to be installed between the engine and cockpit.

25 May: First flight of the only Spitfire delivered to a foreign power before the outbreak of war, the 251st Mk.I off the line (with Class B registration N21) was exported to France. Never allocated a British serial number, the Spitfire was painted in the standard RAF camouflage of the time and later carried the French identity 01. It was delivered on 18 July 1939.

June: Only three months before the outbreak of war there was talk that Spitfire production would be phased out at Supermarine in March 1940 when existing contracts were filled. An Air Ministry memo mentions the Bristol Beaufighter, Westland Lysander or Westland Whirlwind as possible projects to fill Supermarine's production lines, leaving Spitfire manufacture to Castle

A busy scene at Eastleigh Airfield in January 1939 as newly built Spitfire I K9842 is readied for flight testing prior to delivery.

Bromwich. No mention was made of the research and development effort that the Supermarine design team might contribute, but the memo did suggest that export orders could be filled after spring 1940, when CBAF comes on line.

August: Only two weeks before the outbreak of war, Spitfires take part in joint British-French air exercises involving a simulated daylight raid on Paris by Bomber Command Wellingtons and then a night raid on Birmingham by French Bloch 210s. Spitfires from 19 Squadron are scrambled to 'intercept' the Blochs but are unsuccessful, the exercise proving that single engined fighters were unsuitable for night fighter operations.

September: A – possibly *the* – vital link in Britain's defence against aerial attack is in place. The 20 Chain Home radar stations stretched from Southampton on the south coast to Newcastle in the north-east, providing coverage for most of the directions from which German attacks could be expected to come. The stations were called Air Ministry Experimental Stations for security reasons.

3 September: Britain declares war on Germany. On this date, 2,160 Spitfires were on order and 306 had been delivered with production running at the rate of about one per day. Of these, 187 were in squadron service, 71 were held by maintenance units, 11 were engaged in trails, one was with a training unit and no fewer than 36 had been written off. The 10 RAF squadrons equipped with Spitfires on the first day of the war were: Nos 19 (Duxford), 41 (Catterick), 54 (Hornchurch), 65 (Hornchurch), 66 (Duxford), 72 (Church Fenton), 74 (Hornchurch), 602 (Abbotsinch), 609 (Yeadon) and 611 (Speke/Liverpool). The last three were Auxiliary Air Force units.

4 September: The Spitfire prototype K5054 written off after a landing accident at Farnborough; it had recorded 151hr 30min service flying time although the overall total has been estimated at around 260 hours. Flown by Flt Lt G S White, the aircraft began to lift its tail after touchdown and flipped onto its back, killing the pilot. The precise cause of the accident was not found as the undercarriage was locked down, the wheels free, there was no binding of the brakes and the elevator trim was correctly set. It has been speculated that White loosened his harness on landing, the anchorage straps in the rear fuselage dropping and fouling the elevator cables and forcing the elevators into the full down position. Before being scrapped, K5054's fuselage was used as mockup for the installation of cameras in early Spitfire photo-reconnaissance models.

6 September: Sadly, the first victims claimed by the Spitfires' guns were two Hurricanes during the so-called 'Battle of Barking Creek'. A fault in one of the Chain Home radar units resulted in controllers sending 74 Squadron's Spitfires in the wrong direction, where they intercepted Hurricanes from 56 Squadron instead of what was thought to be an incoming German raid.

24 September: The 1,175hp (876kW) Merlin XII (for the Spitfire II) begins flight testing in Spitfire I K9788. This was the first Merlin to use 100 rather than 87 octane fuel, the switch to the higher octane fuel all part of the ongoing quest for improved performance. In combination

THE HIGH SPEED SPITFIRE

First flown on 10 November 1938, the Type 323 High Speed Spitfire was converted from the 48th production Spitfire I (K9834) and intended to capture the World's Landplane Speed record for Britain.

Allocated the Class B marking N17, the aircraft featured a highly boosted (+28lb, more than four times the normal limit) Merlin II Special producing about 2,100hp (1,566kW), reduced wing span, low drag canopy, reduced fuel capacity, clipped wings, highly modified cooling system and tailskid. The fuel was a mixture of petrol (20%), benzol (60%) and methanol (20%), and the engine ran at 3,200rpm, 200rpm more than standard. The propeller was a Watts fixed pitch, four bladed wooden unit and loaded weight was 5,490lb (2,490kg).

Testing revealed a top speed of 408mph (656km/h) at 3,000ft (915m), equivalent to about 400mph (644km/h) at the 200ft (60m) altitude required for a new record to be recognised.

When the idea was first mooted in mid 1937 Howard Hughes held the record of 352.4mph (567.1km/h) in his H-1 aircraft, this speed about 62mph (100km/h) more than the prototype Spitfire was capable of reaching at sea level. The Speed Spitfire was intended to beat this mark (which it did) but the situation changed radically in March 1939 when the Heinkel He 100 V8 set a new record of 464mph (747km/h) which was in turn increased to 469mph (755km/h) by the Messerschmitt Me 209V1 the following month. Both speeds were well outside any Spitfire's capability.

The High Speed Spitfire project was therefore dropped and the aircraft reverted to mostly standard features including fitting a Merlin XII engine as installed in the production Spitfire II. It was used as a hack by the Photographic Reconnaissance Unit (PRU) and survived the war to be scrapped in 1946.

The High Speed Spitfire in its original form (top) and as it appeared in March 1942, operated by the Photo Reconnaissance Unit and converted to more or less Mk.I standards. By then, the aircraft carried the military serial K9834.

A lineup of 611 Squadron Spitfire Is at Duxford in the second half of 1939. Note the three bladed propellers which were by then standard equipment on the Spitfire. (via Neil Mackenzie)

with a Rotol propeller, the new engine and fuel gave the Spitfire better performance at all altitudes as higher boost pressures could be used without the risk of detonation. The 100 octane fuel couldn't be immediately introduced on a widespread basis because at first it all had to be shipped from the USA and German submarine activity in the Atlantic sunk large numbers of tankers. The situation improved when refining was undertaken in Persia and the fuel became readily available from the northern hemisphere spring of 1940. It was also used in the Spitfire I's Merlin III engine with good effect.

16 October: The first *Luftwaffe* attack on mainland Britain and first blood for the Spitfire. Aircraft from 602 and 603 Squadrons intercept nine *Luftwaffe* Junkers Ju 88s attacking naval targets in the Firth of Forth. One

Ju 88 was shot down by 603 Squadron's Flt Lt Pat Gifford and another was shared by 602 Squadron's Flt Lts A McKeller and G Pinkerton. These were the first enemy aircraft destroyed by Fighter Command in WWII.

November: The first mention of the possibility of a Griffon powered Spitfire made by the Ministry of Aircraft Production. After deciding to drop the idea, it was quickly reinstated, Supermarine issuing a specification for a Griffon-Spitfire the following month. This led to the Spitfire IV prototypes.

18 November: The first operational Spitfire photo-reconnaissance sortie performed by Flt Lt Maurice Longbottom in Spitfire I N3071 with armament removed

16 November 1939 and Flt Lt Maurice 'Shorty' Longbottom prepares to take Spitfire I N3071 on the type's first photo-reconnaissance mission over the German city of Aachen. The Spitfire's armament was removed and the gun bays filled with F.24 cameras instead.

PRE WAR EXPORT PROSPECTS

Not surprisingly, the Spitfire created a great deal of interest among potential customers before the war, with 13 countries stating requirements for just under 300 aircraft by mid 1939. Britain's needs, the relatively slow build up of production and the circumstances of some countries as war was declared obviously overrode any export possibilities with the result that only France received a single Spitfire in 1939.

In November 1938 Britain's Foreign Office drew up a list of the countries that wanted Spitfires in order of priority, and they are listed in that order along with the quantity being discussed: France (at least 3), Belgium (up to 60), Estonia (12), Turkey (up to 60), Romania (10), Portugal (15), Switzerland (15), Yugoslavia (12), the Netherlands (up to 36), Greece (up to 36), Bulgaria (12), Iran (24) and Lithuania (12). Of these, France, Switzerland, Yugoslavia, Belgium and the Netherlands had also been negotiating for licence production of the aircraft.

The list of potential customers grew further as 1939 progressed, with Finland, Norway and China adding their names while even Japan requested a single example – it doesn't require much imagination to envisage a copy appearing from a Japanese factory a few months after delivery!

Apart from the French aircraft (which was the 251st Mk.I off the line), some production slots were allocated including Estonia (nos 260 and 261 but cancelled when a non aggression pact with Germany was signed), and two of Turkey's Spitfires were shipped in May 1940 but not delivered.

Mk.I PRODUCT IMPROVEMENT

The first few production Spitfire Is were used largely for trials and development, testing many of the features which would subsequently appear on aircraft issued to the squadrons.

The original de Havilland fixed pitch and two bladed propeller was replaced with a two-position, three bladed propeller from the same manufacturer from the 74th aircraft, and from number 175 (mid 1939) the Merlin III engine replaced the similarly rated Merlin II but with a shaft capable of accepting either de Havilland or Rotol three bladed, constant-speed propellers. The operational limitations imposed on aircraft fitted with the two-position DH propeller led to a rushed modification programme between June and August 1940 which saw those aircraft so equipped have the DH 'Hydromatic' constant-speed unit (based on a Hamilton-Standard design) fitted.

The domed canopy first appeared with the eighth production Mk.I (K9794) – this resulting from complaints that pilots were hitting their heads with the original 'flat' design – while experience in the Battle of Britain saw numerous detail modifications introduced, among them increased armour protection including the fitting of armoured glass to the windscreen, protection for the upper fuel tank (immediately in front of the pilot) and the installation of steel plating behind the pilot's seat. The introduction of Identification Friend or Foe (IFF) equipment from early 1940 and the easily identified 'blip' it produced on British radar screens greatly helped Fighter Controllers during the Battle of Britain.

and a vertical F.24 camera installed in the gun bay of each wing. Departing from Seclin in northern France, Longbottom attempted to photograph the German city of Aachen and its defences from 33,000ft (10,060m) but navigation difficulties resulted in him returning with images from south of the target. Four days later, Longbottom successfully completed the mission. With the introduction of improved PR Spitfire Is, the original two aircraft were retrospectively designated Spitfire Mk.I PR Type A.

20 November: The first Spitfire 'kill' over south-east England when three 74 Squadron aircraft intercepted a Heinkel He 111 which had crossed the Sussex coast. The He 111 was badly damaged and came down in the sea off East Anglia.

21 November: A RN Fleet Air Arm pilot test flies a Spitfire (Mk.I R6718). Royal Navy interest in a 'Sea Spitfire' had revived after a year long hiatus, prompted by a severe shortage of its own fighters. A study into the feasibility of fitting an arrester hook to the Spitfire was requested after the flights, but Supermarine had already test flown a hooked Spitfire on 16 October and presented drawings to the FAA a few days later. Information on the possibility of a folding wing Spitfire was also requested.

12 December: The British Air Staff decides that development of the Spitfire should have priority over the Hawker Hurricane, although the latter was fitted with the more powerful Merlin XX engine which was also intended for the proposed Spitfire III.

15 December: The Fleet Air Arm writes to Supermarine investigating the possibility of quickly building 50 new Spitfires with folding wings and arrester hooks. Supermarine responded the following month with drawings of the Griffon powered Type 338 with the wings breaking at a point just outboard of the undercarriage and swivelling back to fold parallel with the fuselage, sitting vertically with the leading edges upwards. Supermarine stated that deliveries of this model could start 11 months from the order being placed, or five months if a normal Spitfire was used as the basis of the design. Various production proposals and delivery schedules were discussed over the next few months.

BIRTH OF THE PR SPITFIRE

In August 1939, Flg Off Maurice Longbottom RAF, submitted a paper to the Air Ministry outlining views on the future of aerial photographic reconnaissance. It called for the use of "....a single small machine relying on its speed, climb and ceiling to avoid detection.... as most fighters have a very good takeoff performance due to their great reserve of engine power, they could be considerably overloaded with further fuel to give an even greater range."

Longbottom presented the case for using the Spitfire I in this role, arguing that with the removal of armament and other equipment, some 450lb (204kg) in weight could be saved and the all up weight could be increased by 480lb (218kg) over standard. Fuel capacity could be trebled and the range increased to 1,500 miles (2,414km) even after cameras and extra tanks had been installed.

The Ministry showed little interest in the idea at first but operational experience with Blenheims and Lysanders in the opening month of the war – both suffered heavy losses on even short range PR missions – quickly encouraged a change of heart. Two Spitfire Is (N3069 and N3071) were released for conversion to PR standards and delivered to the secret Heston Flight reconnaissance unit commanded by Wng Cdr Sidney Cotton.

An Australian, Cotton was largely responsible for establishing the Heston Flight and had surreptitiously photographed various parts of Germany – including the entire Ruhr Valley, aerodromes and military installations – in the months leading up to the war using a pair of Lockheed 12As. Cotton flew the aircraft over Europe as a private businessman representing the 'Aeronautical Research and Sales Corporation', which had been set up especially as cover for his activities.

Both the Spitfires were fitted with a 5in (12.7cm) focal length vertical camera in each wing. No additional fuel capacity was provided but they were 'cleaned up' (gun ports sealed, joints filled etc) to provide an extra 10-15mph (16-24km/h) of speed. Longbottom was posted to the unit (now as a Flt Lt) and flew the first mission on 18 November 1939.

The Heston Flight was renamed No 2 Camouflage Unit at around the same time (to explain the unique bluish-green 'Camotint' colour scheme originally applied to the aircraft) but in early 1940 became the Photographic Development Unit (PDU) and in July of the same year the definitive Photographic Reconnaissance Unit (PRU). Camotint fell from favour as it was found to be unsuitable for high altitude operations, replaced by the universally known 'PR Blue'.

Jeffrey Quill (left) and a high level French contingent during a visit to Southampton and a Spitfire demonstration in May 1939. A Spitfire I – the 251st off the line – was delivered to France later in the same month, the only one to be exported before the outbreak of war.

Spitfire I L1090 was shipped to Canada in May 1940 for comparative testing against a USAAC Curtiss XP-40 (which can just be seen on the far left). The trials were conducted at the RCAF's Uplands base. The Spitfire carries the small roundels applied to aircraft built in mid 1939 but the small serial number on the fin is non standard. L1090 returned to Britain after the trials.

January: Production of 30 Spitfires with two 20mm Hispano cannon authorised and delivered from June 1940. The cannon were prone to jamming and were subsequently supplemented by four 0.303in machine guns housed in the 'B' wing. The new variant was designated Mk.IB and the standard model with eight machine guns retrospectively designated Mk.IA. Operational trials were conducted by 19 Squadron.

16 January: Delivery of the first Spitfire Mk.I PR Type B with fuselage fuel tank and longer focal length vertical cameras.

22 February: Flg Off George Proudman and Sqn Ldr Douglas Farquhar of 602 Squadron based at Drem in Scotland bring down a Heinkel He 111. The action was significant because Proudman was flying the prototype Spitfire IB (L1007), the only cannon armed Spitfire in RAF service at the time. Proudman managed to get off 31 rounds before the guns jammed, sufficient to claim part of the first Spitfire cannon 'kill'.

March: Rolls-Royce begins work on the Merlin 60 series engine with two-speed/two-stage supercharger.

March: Delivery of first Spitfire I PR Type C with blister tank under the port wing, balanced by two cameras in a similar blister under the starboard wing.

15 March: First flight of Spitfire III prototype (N3297). Despite 1,000 of this first major development of the aircraft being ordered, only two prototypes were built and the programme was cancelled. N9327 later became the prototype for the Spitfire IX, while the second Mk.III (W3237, converted from a new Mk.V airframe) first flew on 4 June 1941. It was used for various trials including extended wing tips, four position flaps as part of the Seafire development programme (the standard Spitfire's flaps were either fully up or fully down) and roll performance tests with standard or clipped wings and different gun installations.

16 March: The First Sea Lord (and soon Prime Minister) Winston Churchill orders cancellation of the folding

A Spitfire I of No 72 Squadron when the unit was most likely based in Scotland in late 1939/early 1940. It moved south later in 1940 when the Battle of Britain was raging. Note the early radio mast. (via Neil Mackenzie)

wing Spitfire for the Royal Navy as other apparently more suitable specialised naval types were under development. It would be another 17 months before the idea was revived, leading to development of a 'Sea Spitfire' – the Seafire.

May: By the beginning of the month the RAF had 19 squadrons of Spitfires on strength: Nos 19 (based at Horsham St Faith), 41 (Catterick), 54 (Hornchurch), 64 (Catterick), 65 (Northolt), 66 (Duxford), 72 (Acklington), 74 (Hornchurch), 92 (Northolt), 152 (Acklington), 222 (Duxford), 234 (Leconfield), 266 (Wittering), 602 (Dyce), 603 (Drem), 609 (Drem), 610 (Prestwick), 611 (Digby) and 616 (Leconfield). These bases changed frequently, more so from this month when Germany launched its *Blitzkrieg* offensives on France, Belgium and the Netherlands. A couple of the Scottish based Auxiliary Air Force squadrons were moved south later in the month to help meet anticipated needs there, 609 redeploying to Northolt and 610 to Biggin Hill.

10 May: The Spitfire squadrons of Nos 10 and 11 Groups are authorised to conduct cross-Channel offensive patrols.

12 May: The first Spitfire cross-Channel offensive patrol involving six aircraft from No 66 Squadron accompanied by six Boulton Paul Defiants of 264 Squadron. The sweep took the aircraft over Holland, damaging a Junkers Ju 88 in the process.

13 May: The Spitfires of No 66 Squadron and Defiants of No 264 Squadron fly a patrol over Holland similar to the one the previous day. On this occasion the Spitfires came up against the Messerschmitt Bf 109 for the first time. The result of the skirmish was one Bf 109, four Junkers Ju 87s, one Spitfire and five Defiants shot down.

13 May: Spitfires from 74 Squadron (Hornchurch) provide cover for the destroyers HMS *Hereward* and *Windsor* which are bringing members of the Dutch Royal Family and Government to exile in England.

PRODUCTION DISPERSAL

Supermarine made substantial use of sub contractors for major Spitfire components right from the start in order to help achieve a reasonable rate of production. It must be remembered that between 1919 and 1936 the company had built a grand total of only 137 aircraft and the placing of large orders for the Spitfire took it into unexplored territory!

Production was undertaken at the company's two major works in the Southampton area – Woolston and Itchen – with Folland, General Aircraft, Pobjoy, Westland and others acting as sub contractors. Flight testing was conducted at Eastleigh Aerodrome.

Following the bombing of the two Supermarine facilities in September 1940, Spitfire parts manufacture was dispersed over 46 small production and 19 support facilities in southern England, these comprising mainly garages and workshops staffed by workers who in most cases had to learn their skills from scratch. Spitfire final assembly lines were established at Eastleigh, High Post, Keevil, Kenley and Aldermaston.

It is well known that the Spitfire was relatively difficult to produce because of some of its structural complexities, especially those associated with the wing leading edges. As a result, production built up slowly with only 45 delivered by the end of 1938. The programme was about six months behind schedule in March 1939, by which time the 300 should have been delivered to the RAF.

The Spitfire III prototype (N3237) first flew on 15 March 1940.

Spitfire I P9372 of 92 Squadron. This unit operated from Biggin Hill at the climax of the Battle of Britain. (via Neil Mackenzie)

14 May: Lord Beaverbrook appointed Minister of Aircraft Production, his task to co-ordinate and speed up the complex business of having industry build lots of aircraft very quickly – especially fighters in 1940. The Canadian Beaverbrook pulled no punches and just three days after his appointment was on the telephone to Lord Nuffield demanding to know why no Spitfires had as yet emerged from the Castle Bromwich factory. According to a witness to the conversation, Nuffield said (somewhat sarcastically and after enduring a tirade from the Minister): "Perhaps you would like me to give up control of the Spitfire factory?", to which Beaverbrook replied – as quick as a flash – "Nuffield, that's very generous of you. I accept!" He then hung up and immediately transferred management of the plant to Vickers-Armstrong. Beaverbrook 1, Nuffield nil! (see separate box story).

16 May: Flg Off Robert Stanford-Tuck (shortly to become one of the RAF's leading aces) and Plt Off Alan

Wright of No 92 Squadron fly their Spitfires to Le Bourget (Paris), escorting the de Havilland Flamingo carrying British Prime Minister Winston Churchill to talks with his French opposite number.

23 May: Spitfire I K9867 of 74 Squadron becomes the first of type to be captured by the Germans after sustaining battle damage and being forced down at Calais.

26 May-3 June: The period of the evacuation of the British Expeditionary Force from Dunkirk sees Fighter Command's sortie rate increase considerably, a total of 2,739 being flown during the nine days. Pilots claimed the destruction of 377 enemy aircraft during that period (later officially reduced to 262) but Spitfire losses were put at 72, nearly one-third of Fighter Command's total force.

June: The first Spitfire IIs leave the Castle Bromwich Aircraft Factory (CBAF) in Birmingham. The Mk.II was the first Spitfire model to be built at this mass production

The Mk.II was the first Spitfire mark to be manufactured at the Castle Bromwich Aircraft factory in Birmingham, from June 1940. This particular aircraft (P7665) of No 65 (East India) Squadron was built in November 1940. (via Neil Mackenzie)

The quest to increase the Spitfire's range was never ending, trials conducted during 1940 including a mock up installation of fixed underwing tanks on Mk.I K9791 (top) in January and a single 30imp gal (136 litres) tank under the port wing of P9565 seven months later. The single tank configuration caused handling problems at high speeds.

facility. The first Mk.II (P7280) was delivered to Boscombe Down for testing on 27 June, preceded by the second aircraft (P7281), which had gone to No 8 Maintenance Unit for equipment fitting before delivery to 611 Squadron at Digby on 27 August.

June: The Turkish Government had ordered 60 Spitfire Is in January 1939 comprising 15 complete aircraft and 45 sets of components for local assembly. All but two of the order was cancelled, but P9566 and P9567 were shipped in June 1940 as the Type 341 carrying the British Class B registrations N22 and N23. They were not delivered and turned up in the Middle East during 1942 wearing the RAF serials HK854 and HK856, these not proper Spitfire serials but within a block reserved for impressed aircraft. Turkey later received Spitfire Vs (in 1944).

June: Start of the rushed programme to replace the de Havilland two-position propellers on many Spitfire Is with constant-speed 'Hydromatic' units from the same manufacturer. The programme was largely complete by August.

18 June: Although not very suitable for night fighting, the Spitfire did nevertheless score a few kills in this role, the first one on this date. While undertaking a night interception of some Heinkel He 111s operating over East Anglia, Flg Off Petra spotted one in the moonlight. Petra and the Heinkel's rear gunner opened fire simultaneously with the result that both aircraft went down. Although badly injured, Petra survived to claim the Spitfire's first night victory.

26 June: First bench tests of the Rolls-Royce Griffon II engine.

July: Delivery of first Spitfire I PR Type F, an interim long range model with blister tanks under each wing plus a rear fuselage tank, resulting in more than double the standard Spitfire I fuel capacity. Cameras were located behind the cockpit and the engine cowling was deepened to accommodate the larger engine oil tank necessary for longer flights. Most earlier Spitfire I PR conversions were modified to this standard.

July: Rolls-Royce and Packard conclude arrangements for licence production of the Merlin in the USA, initially covering 1,500 engines with options on a further 10,000.

10 July: Official start of the Battle of Britain, at which time 19 of Fighter Command's 50 single seat fighter squadrons were equipped with 292 Spitfires of which 199 were serviceable.

August: Westland receives its first order to build Spitfires – 300 Mk.Is which are completed as 50 F.IAs, 140

British Prime Minister Winston Churchill chats with Castle Bromwich's chief test pilot Alex Henshaw. A record breaking pilot before the war, Henshaw was probably the world's most experienced Spitfire pilot during the war, reportedly having flown one in ten of all the Spitfires built – that's over 2,000 individual aircraft.

F.VBs and 110 F.VCs between July 1941 and September 1942. Westland went on to manufacture 685 Spitfires and 1,472 Seafires.

August: Rolls-Royce begins work on adapting the Bendix-Stromberg 'injection' carburettor to the Merlin XX and sends drawings of Merlin XX to Packard in the USA to form the basis of the Packard-Merlin 28/V-1650-1 engine. Rated at 1,300hp (969kW) for takeoff, the V-1650-1 powered the Avro Lancaster III, Curtiss P-40F Kittyhawk II and Canadian Hurricane.

13 August: Start of the Battle of Britain 'proper' with the launch of Operation *Adler Angriff* (Eagle Attack), the *Luftwaffe* air offensive against Britain. The initial objective was to nullify RAF Fighter Command by attacking its airfields and drawing the fighters into combat.

18 August: It becomes official policy that wherever possible, Spitfire squadrons should engage *Luftwaffe* fighters and Hurricanes should intercept bombers. This was easier said than done in the heat of battle, especially when a single RAF squadron often found itself facing a large force of enemy aircraft. By the end of the Battle of Britain, Spitfires had nevertheless accounted for more Messerschmitt Bf 109s than had Hurricanes.

20 August: Spitfire I X4257 flies as the prototype for the proper 'B' wing Spitfire with two 20mm Hispano cannon and four 0.303in Browning machine guns. The original Mk.I conversions with just the two cannon but no machine guns had proved problematical in service due to the unreliability of the Hispanos and a compromise was sought. Mk.I P9504 was on hand having had the two cannon fitted but with four of the original eight machine guns still in place. The combination proved to be successful, leading to the manufacture of a new set of wings with mixed armament for installation on X4257 – the first 'B' wing. Another seven Spitfire Is were quickly con-

verted to the new standard and after testing, the cannon jamming problems – largely due to the cartridge ejection chute design – were resolved. Machine gun armed Spitfires were retrospectively dubbed Mk.IA and those with the mixed armament Mk.IB.

24 August: The *Luftwaffe* bombs London and Churchill orders the first of several retaliatory raids the next night. This leads to a switch in Germany's tactics (see 7 September).

27 August: No 611 Squadron at Digby becomes the first to equip with the Spitfire II. Nos 19 (Fowlmere), 74 (Coltishall) and 266 (Wittering) also receive the new model over the next month.

September: The trial installation of a 30imp gal (136 l) external fuel tank on the leading edge of the port wing tested on a Spitfire I at the A&AEE. The concept was planned to be used operationally on Spitfire IIs in July 1941 when five squadrons were scheduled to provide an escort for Operation *Sunrise*, the attack on the German capital ships *Scharnhorst*, *Gneisenau* and *Prinz Eugen* in Brest Harbour. However, the aircraft became operational too late to participate.

September: The first of two Spitfire Mk.I PR Type D prototypes (P9551 and P9552) with integral fuel tanks in the wing leading edges plus a rear fuselage tank delivered. With 2½ times the standard Spitfire I fuel capacity, the Type D was capable of ranging far and wide, including to Berlin. The Type D formed the basis of the production PR.IV utilising a Spitfire V engine/airframe combination.

7 September: The Battle of Britain enters a new phase when the *Luftwaffe* switches the main thrust of its attacks from the airfields to major cities, especially London. Airfields, factories and other strategic targets continued to be attacked, however. The move was a

CASTLE BROMWICH

The Spitfire II was the first version of the fighter to emerge from the Castle Bromwich Aircraft Factory (CBAF) in June 1940. This large new works had been built in the Birmingham area for aircraft production under the direction of the Nuffield organisation, better known for its manufacture of Austin cars.

The factory was proposed in May 1938 when the Air Minister, Sir Kingsley Wood, approached Lord Nuffield (William Morris) on the matter. Lord Nuffield agreed – on the condition a large order was placed – and purchased a 1,414 acre (572 hectare) site belonging to the Birmingham Corporation and intended for housing for £1,000 per acre. The Minister turned the first sod on 1 July 1938 and the first order (for 1,000 Spitfire IIs) was formally placed on 12 April 1939.

There were many teething problems at the factory, attributable to a number of causes and resulting in delays to Spitfire production. The causes included inexperienced management, a lack of skilled labour, the structural complexity of the Spitfire, the effects of enemy air raids and – unbelievably in those desperate days – union disputes.

The shortage of skilled labour was ironic because the government had originally wanted the factory to be in Liverpool to help alleviate an unemployment problem. Lord Nuffield had argued that the lack of skilled workers in Liverpool would cause delays and the workforce in Birmingham – which was experienced in motor vehicle production – could be trained much more quickly.

Under orders from Lord Beaverbrook – immediately on his appointment as Minister for Aircraft Production – management of the CBAF was transferred to Vickers-Armstrong in May 1940 and senior staff brought in from Supermarine to get the Spitfire line moving. The first Spitfire (Mk.II P7280) emerged from the CBAF line in June 1940 and 1,300 had been produced by the end of 1941. The last CBAF Spitfire was Mk.22 PK614, flown on 30 November 1945.

Once the problems had been sorted out, CBAF became the major Spitfire production source, building 11,989 or just over half the production total between 1940 and the end of the war.

CBAF concentrated on the 'mainstream' Spitfire versions – Mks.II, V, IX, XVI – as well as the later Mks.21, 22 and Seafire 45 – while Supermarine built the lower volume models. Westland and Cunliffe-Owen Aircraft also contributed to the Spitfire/Seafire tally.

Apart from Spitfires, CBAF was also responsible for the manufacture of 300 Avro Lancasters in 1943-45.

The Spitfire production line at Castle Bromwich. The facility was responsible for the manufacture of more than half of all the Spitfires and Seafires built.

retaliatory one on Germany's part in response to some sporadic and largely ineffectual raids on Berlin which had taken place during the last week of August – and it gave Fighter Command a little breathing space as the raids on its airfields reduced.

15 September: The day regarded as the climax of the Battle of Britain, the *Luftwaffe* launching 200 bombers against London in two waves. Attacking the first wave left Fighter Command with no squadrons in reserve; the second wave came through two hours later giving the squadrons time to refuel and rearm – any earlier and the RAF would have been caught on the ground. Fifty-two German bombers were shot down during the actions at a cost of 29 RAF fighters – seven of them Spitfires and the remainder Hurricanes. Hitler indefinitely postponed Operation *Sea Lion* – the invasion of England – the next day.

24 September: The first of two heavy *Luftwaffe* raids on Supermarine's Southampton facilities, this one on the Itchen works and leaving 90 workers dead when an air raid shelter received a direct hit.

26 September: Supermarine's Itchen factory attacked for the second time in three days and the Woolston works gutted by what has been described as a "textbook carpet bombing raid" by Heinkel He 111s. The lives of 135 workers were lost plus civilians in the surrounding area. By now Spitfire production was to some extent dispersed but there were disruptions to the schedule with deliveries in October dropping to 151 aircraft compared to 268 in September. The *Luftwaffe* also bombed the Westland factory at Yeovil late in the month, just as it was starting to tool up for Spitfire production.

27 September: The most expensive day of the Battle of Britain in terms of Spitfire losses with 15 destroyed defending attacks on London, Filton (Bristol) and in the north.

1 October: Another change in the *Luftwaffe's* tactics in the face of unacceptable bomber losses, with increasing use of Messerschmitt Bf 109 fighter-bombers operating at higher altitudes rather than heavily escorted bomber formations.

19 October: Formation of the RAF's first Eagle Squadron (No 71) at Church Fenton in Yorkshire, established for American volunteers who wished to help in the fight against Germany. The squadron was originally equipped with Hurricanes but switched to the Spitfire II (quickly followed by the Mk.V) in August 1941. Two other Eagle Squadrons were formed in 1941 as Nos 121 and 133, the three units becoming the 334th, 335th and 336th (respectively) Fighter Squadrons of the Eighth Air Force's 4th Fighter Group in Britain from September 1942.

24 October: 1,000 Spitfire IIIs ordered from Castle Bromwich but following cancellation of the programme, these were built as Mk.VBs instead.

31 October: The Battle of Britain effectively ends, the *Luftwaffe* switching to an increasing number of night raids.

25 November: Air Marshal Sholto Douglas appointed Commander-in-Chief of RAF Fighter Command.

20 December: In order to 'test the waters' over occupied France, two Spitfires of No 66 Squadron flown by Flt Lt G Christie and Plt Off C Brodie departed Biggin Hill, crossed the English Channel and then flew low across France until reaching a point north of Paris. They shot at targets of opportunity on the way home (including strafing buildings at Le Tourquet airfield) in what was the first offensive fighter sortie over enemy territory since the Battle of France six months earlier. The first official offensive fighter sweep was conducted the following month.

26 December: Spitfire IA K9788 flies with a Merlin 45 engine installed, in effect the prototype for the Spitfire V.

THE BATTLE OF BRITAIN

It is not the intention here to retell the story of the Battle of Britain but rather to present some facts and figures pertaining to the Spitfire's involvement in this most crucial of battles, one which determined not only the course of the war with Germany but Britain's (and Europe's) very future.

Generally speaking, the Battle of Britain comprised three phases. The preliminaries took place over the two month period starting in the first week of June 1940 when *Luftwaffe* activities were directed mainly against the ports on England's south coast. The official start of the Battle is regarded by the British as being 10 July 1940 – when the *Luftwaffe* began attacking shipping in the English Channel – and a week earlier by Germany.

The Battle proper is considered to have started on 13 August, the day Reichsmarschall Hermann Goering declared as *Adlertag* – the 'Day of Eagles'. On that day, the *Luftwaffe* began a series of large, sustained and

"Tally ho!" Spitfires peel off to once again mix it with the Luftwaffe during the Battle of Britain. Such set formations were found to be a liability during the Battle and the loose formations used so effectively by the Luftwaffe were adopted.

Heinkel in trouble. An He 111 getting some close – and probably terminal – attention from a Spitfire.

well escorted attacks on RAF airfields and other military targets.

After a period of about three weeks, during which RAF Fighter Command suffered heavy losses and appeared to be in trouble, Goering switched the main thrust of his attacks away from military targets and on to major cities, particularly London from 7 September. This move gave the RAF some much needed breathing space and ultimately helped it in its quest to repel the *Luftwaffe*, although this did not mean any slackening in the intensity of the fighting.

The climax of the Battle was reached on 15 September 1940 – now commemorated as Battle of Britain Day – when the *Luftwaffe's* last major offensive was launched and repelled by Fighter Command. After that, Germany had to concede it had not achieved the air superiority over Britain that Goering had promised and invasion plans were indefinitely postponed.

The raids continued into the following month, the last day of the Battle of Britain being officially recognised as 31 October. The *Luftwaffe* then switched a new phase of raids on Britain, the night *Blitz*.

RAF Fighter Command was headed by Air Marshal Sir Hugh Dowding and organised into four Groups for the defence of the British Isles: 10 Group (covering south-west England and most of Wales); 11 Group (the

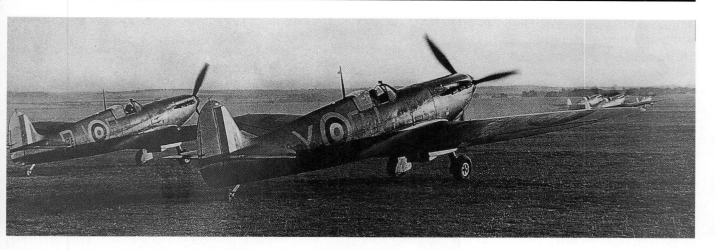

A 616 Squadron 'scramble' from Kenley in July 1940.

heavily populated south-east of England including London); 12 Group (the midlands); and 13 Group (the north of England and Scotland). 11 Group – brilliantly commanded by Air Vice-Marshal Keith Park who worked closely and effectively with the inspirational Dowding – took the lion's share of the workload and therefore had the greatest number of squadrons allocated to it.

No 10 Group was established relatively late in the day on 21 July 1940 – until then it was part of 11 Group.

Spitfire Order of Battle 1 July 1940

No 10 Group: No Spitfire squadrons established.
No 11 Group: Eight Spitfire squadrons (Nos 54, 64, 65, 74, 92, 234, 609, 610) with 91 serviceable and 38 unserviceable aircraft (also 283 Hurricanes).
No 12 Group: Five Spitfire squadrons (Nos 19, 66, 222, 266, 611) with 43 serviceable and 29 unserviceable aircraft (also 48 Hurricanes).
No 13 Group: Six Spitfire squadrons (Nos 41, 72, 152, 602, 603, 616) with 64 serviceable and 28 unserviceable aircraft (also 128 Hurricanes).
Total Spitfire strength: 198 serviceable, 95 unserviceable (plus 459 Hurricanes).

Spitfire Order of Battle 1 August 1940

No 10 Group: Four Spitfire squadrons (Nos 92, 152, 234, 609) with 43 serviceable and 14 unserviceable aircraft (also 50 Hurricanes).
No 11 Group: Six Spitfire squadrons (Nos 41, 64, 65, 74, 266, 610) with 70 serviceable and 26 unserviceable aircraft (also 216 Hurricanes).
No 12 Group: Four Spitfire Squadrons (Nos 19, 66, 222, 611) with 52 serviceable and 13 unserviceable aircraft (also 51 Hurricanes).
No 13 Group: Six Spitfire squadrons (Nos 54, 72, 602, 603, 610, 616) with 64 serviceable and 22 unserviceable aircraft (also 149 Hurricanes).
Total Spitfire strength: 229 serviceable, 75 unserviceable (plus 466 Hurricanes).

Spitfire Order of Battle 1 September 1940

No 10 Group: Four Spitfire squadrons (Nos 92, 152, 234, 609) with 47 serviceable and 18 unserviceable aircraft (also 61 Hurricanes).
No 11 Group: Six Spitfire squadrons (Nos 54, 72, 222,

602, 603, 616) with 75 serviceable and 22 unserviceable aircraft (also 219 Hurricanes).
No 12 Group: Five Spitfire squadrons (Nos 19, 66, 74, 266, 611) with 52 serviceable and 25 unserviceable aircraft (also 62 Hurricanes).
No 13 Group: Two Spitfire squadrons (Nos 41 and 64) with 26 serviceable and nine unserviceable aircraft (also 133 Hurricanes).
Total: Spitfire strength: 200 serviceable, 74 unserviceable (plus 475 Hurricanes).

Group Bases

10 Group Spitfire Bases: Exeter, Middle Wallop, Pembrey, Warmwell.
11 Group Spitfire Bases: Biggin Hill, Croydon, Gravesend, Hornchurch, Kenley, Northolt, Pembrey, Rochford, St Eval, Tangmere, Westhampnett.
12 Group Spitfire Bases: Coltishall, Digby, Fowlmere, Kirton-in-Lindsay, Ternhill, Wittering.
13 Group Spitfire Bases: Acklington, Catterick, Church Fenton, Drem, Leconfield, Turnhouse, Usworth.

Spitfire Losses and Deliveries
July-October 1940

	July	Aug	Sept	Oct	Total
Delivered	83	245	268	151	747
Destroyed	27	136	133	65	361
Damaged	51	110	133	58	352

One of the more famous images from the Battle Britain – Spitfire Is of 610 (County of Chester) Squadron in battle formation. 610 was based at Biggin Hill in July, August and part of September 1940 and was therefore involved in the thick of the fighting. (via Neil Mackenzie)

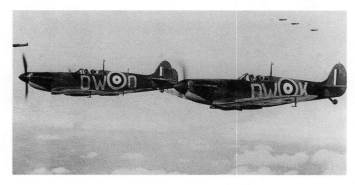

January: First production Merlin 45 (for the Spitfire V) delivered.

10 January: The first official 'Circus' offensive mission into Europe flown, involving three squadrons of Spitfires, three of Hurricanes and one comprising six Blenheims. An ammunition dump in the Foret de Guines was bombed. A Circus operation was designed to draw enemy fighters into combat and was one of several different types of offensive sweep, although they were usually reluctant to do so, especially when the RAF fighters were operating alone (see separate box).

February: Specification F.4/41 issued, written around Supermarine's proposals for a Griffon powered Spitfire. Points within the specification included maximum speed not less than 410mph (860km/h); ceiling not less than 39,000ft (11,885m); armament six 20mm cannon, or two 20mm cannon and eight 0.303in machine guns, or 12 0.303in machine guns.

17 February: The Air Ministry requests development of a Spitfire with a pressurised cabin. A Spitfire I (R7120, first flight 12/2/41) converted to a Mk.V was taken from the production line and partially dismantled to allow the fitting of cabin pressure bulkheads, a sealed canopy and Marshall blower. The aircraft was initially fitted with a Merlin 47 (as used in the production Spitfire VI) but first flew with a Merlin 61 installed on 24 August 1941. This was intended to be the Spitfire VI production standard but as the 60 series Merlins would not be available in the planned timescale, it was decided that the Mk.VI would have a Merlin 47.

March: Deliveries of new Spitfire Vs begins, initially

A 'scramble' for the pilots of No 340 Squadron, a Free French unit established in Scotland in late 1941 flying Spitfire IIAs. (via Neil Mackenzie)

from Supermarine and Westland; production at Castle Bromwich started in June. These were preceded by Mk.Vs which had been converted from Mk.Is on the production line.

14 March: The first Allied aircraft to overfly Berlin in daylight successfully completes its sortie: Spitfire PR Type F X4712 flown by Sqn Ldr Patrick Ogilvie of 3 PRU takes 95 photographs of the German capital in a mission flown from Alconbury.

April: Rolls-Royce begins bench testing of the Merlin 60 series with two-speed/two-stage supercharger.

April: The first of the Commonwealth units formed in the United Kingdom to operate the Spitfire are formed. The Royal New Zealand Air Force's No 485 Squadron was the only 'kiwi' unit to fly Spitfires in RAF Fighter

The Spitfire V was the next major production variant, entering service in early 1941. AA937 was a Mk.VB built by the parent company in November 1941. It is shown here a year later with clipped wings while conducting trials with the Air Fighting Development Unit at Boscombe Down. (via Neil Mackenzie)

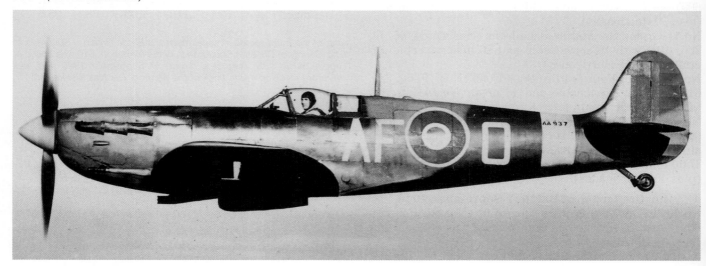

Spitfire I AR212, the first Spitfire from Westland's production line and built in July 1941. Westland continued manufacturing the Mk.I until the end of 1941, nearly a year after the parent company had phased it out. Delivered direct to No 57 Operational Training Unit, AR212 was written off after a collision with Spitfire II X4605 (converted from a Mk.I) in March 1942. (via Neil Mackenzie)

Command. Also in April, the first Royal Australian Air Force Spitfire squadron operating as part of Fighter Command – No 452 – was established using Mk.Is. Three other RAAF squadrons – Nos 451, 453 and 457 – followed between then and 1943.

May: No 92 Squadron becomes the first unit to receive Spitfire Vs in its new production (rather than converted) form.

May: The first of 11 Royal Canadian Air Force squadrons to operate Spitfires in Britain under RAF command – No 403 – is formed, initially on the Mk.I. All but one of these squadrons operated from Britain before moving into the European Theatre of Operations. The exception was 417 Squadron which went from the UK to the Western Desert and Italy. Three of Canada's home based squadrons were moved to Britain in 1944 in preparation for D-Day, three of them equipped with Spitfires and renumbered as Nos 441, 442 and 443. They served with the 2nd Tactical Air Force.

12 May: The Ministry of Aircraft Production rules that all new production Merlin 46 and 47 engines (with larger diameter supercharger impeller for higher altitude operations) will have negative 'g' carburettors. The Merlin 46 was fitted to some Spitfire Vs and the 47 with a cabin blower drive was for the lightly pressurised Mk.VI.

23 May: First flight of Spitfire VB W3248, a trials aircraft which later became the prototype 'clipped wing' Spitfire with the tips removed and replaced with a wooden plug. W3248 was also involved in external mirror, cockpit sight parameter, clear view

and cockpit misting trials. In 1944 it was used in comparative trials with the de Havilland 'Spider Crab', codename for the prototype Vampire jet fighter.

26 May: An order placed with Supermarine for two Griffon powered Spitfire IV prototypes, serial numbers DP845 and DP851.

June: A review of Spitfire development results in the decision being taken to test the new Merlin 60 series engine with two-speed/two-stage supercharger in the Spitfire V. The eventual result was the Mk.IX.

4 July: First flight of the prototype Spitfire VI (X4942), a converted Mk.V in turn converted from a Mk.I.

August: Bench testing of 1,565hp (1,167kW) Merlin 61 for Spitfire VII/VIII/X/XI begins.

Photo-reconnaissance became an increasingly important role for the Spitfire in 1941. This PR Type G (later PR.VII) conversion of a Mk.VA was unusual among Spitfire photo-recce variants in retaining an armament of eight machine guns. It is photographed in the service of No 1416 Flight at Hendon, a tactical reconnaissance unit.

DP845, the first of two prototype Spitfire IVs and the first to fly with a Griffon engine. It recorded its maiden flight on 27 November 1941.

The prototype Spitfire III (N3297) was fitted with a Merlin 61 engine, normal span wings and a four bladed propeller to become in effect the first Spitfire IX. It first flew in its new guise on 27 September 1941.

August: First Packard-Merlin V-1650-1 bench tested.

20 August: A Russian pilot flies a Spitfire for the first time at Duxford. The Soviets did not specifically request Spitfires for more than a year, after the September 1942 arrival in North Russia of some RAF photo-reconnaissance examples which were there to support RAF Handley Page Hampden bombers. The Hampdens had been temporarily deployed to the area to protect a particular convoy. The direct supply of Spitfires to the Soviet Union would start in March 1943 (see below).

23 August: The first production contract for Griffon powered Spitfires placed with the Castle Bromwich Aircraft Factory (three months before the prototype had flown) covering 750 of the proposed initial model, the Mk.IV. The contract was subsequently changed and the aircraft completed as Merlin powered Spitfire VBs and VCs.

27 September: The prototype Spitfire III (N3297) flies with a Merlin 61 engine, standard span wings and a Rotol four bladed propeller installed, in effect becoming the prototype Spitfire IX. Coincidentally, this was also the day the Spitfire V first met the superior Focke-Wulf 190

in combat, this and subsequent encounters quickly establishing an urgent need for the Spitfire IX.

October: Following several months of further discussion about naval Spitfires, Supermarine is requested to carry out investigations into fitting the Spitfire with catapult launching and deck arresting gear. Mk.VB AB205 was allocated to the task, first flying as in effect the prototype Seafire IB conversion on 6 January 1942.

October: Introduction of the 'C' or 'Universal' wing to Spitfire V production with the ability to carry an armament of either eight machine guns, four 20mm cannon or two cannon and four machine guns. The vast majority had the latter installation and the eight machine guns layout was dropped before production began.

October: Official testing of the Spitfire III prototype re-engined with a Merlin 61 reveals a considerable improvement in performance including a maximum speed of 422mph (679km/h) at optimum altitude.

November: The decision is taken to fit metal skinned (rather than fabric) ailerons to all Spitfire Vs, these improving manoeuvrability at high speeds. It was a slow process – by mid 1942 they had only just started to be fitted to aircraft on the production line and there were large numbers of aircraft needing to have them retrofitted. Of interest is the fact that when the Americans began operating Spitfires they replaced the fabric covered ailerons in their Mk.Vs with plywood skinning. RAAF crews assembling Spitfire Vs as they arrived in Australia from late 1942 discovered three variations of aileron construction: metal frame/fabric covered, wooden frame/fabric covered, and wooden frame/plywood covered. Rudders also varied, being either metal or wooden framed with fabric covering in both cases.

13 November: British Prime Minister Winston Churchill orders a halt to all but essential fighter operations over Northern Europe due to heavy losses in the many large scale offensive sweeps which had been taking place throughout the year. The appearance in August of arguably Germany's best fighter – the Focke-Wulf 190 – was a significant factor.

The situation wasn't helped by exaggerated claims from 11 Group's commander, AVM Trafford Leigh-Mallory, who had often been at loggerheads during the Battle of Britain with Dowding and 11 Group's then chief Keith Park. Leigh-Mallory claimed that his fighters had destroyed 437 German fighters on offensive sweeps between mid June and early September 1941 with another 182 probables. The real figure was 128 destroyed and 76 damaged while Fighter Command lost 194 pilots during the same period – not a good ratio. The exaggerated claims continued, the RAF claiming 731 enemy aircraft destroyed between June and December 1941, the real

CIRCUSES and RHUBARBS

In early 1941 the RAF's mood changed from defence to offence following the Battle of Britain. Early daylight offensive sweeps into Europe involving fighters were little more than cheeky displays of defiance, but they soon became organised and effective incursions into enemy occupied territory.

There were several variations of these offensive operations, each with a specific purpose:

Circus: A bomber or fighter-bomber operation heavily escorted by fighters and intended mainly to entice enemy fighters into the air.

Sweep: General term covering fighters flying offensive missions over enemy territory or the sea, with or without accompanying bombers.

Rodeo: Fighter sweep over enemy territory without accompanying bombers.

Ramrod: Similar to a circus but with the aim of destroying a specific target.

Rhubarb: Freelance fighter operations on a small scale attacking targets of opportunity. These were often conducted in bad weather so as to introduce an element of surprise.

The first Spitfire V with the large Vokes tropical filter flew in December 1941, the filter making a substantial difference to engine life in dusty operating environments. This tropicalised Mk.VC (JG867) was built at Castle Bromwich in late 1942. (via Mike Kerr)

The high altitude Spitfire VI with pressure cabin and extended span wings first flew in July 1941 and the first production aircraft appeared in December.

figure being only 103 in combat. Fighter Command admitted to no fewer than 411 losses.

After a period of training, general operations recommenced in April 1942 but heavy losses to the Fw 190 continued, Fighter Command losing 335 aircraft over the next three months, most of them Spitfire Vs. Operations were again curtailed in June 1942, but by then the answer to the Fw 190, the Spitfire IX with Merlin 60 series engine, was just entering squadron service.

27 November: First flight of the prototype Spitfire IV (DP845) with 1,735hp (1,294kW) Griffon IIB. The second prototype (DP851) flew the following month and by March 1942 both were being referred to as the Mk.XX to avoid confusion with the Spitfire PR.IV. The Spitfire IV never went into production and both aircraft were converted to prototypes for other Griffon Spitfire versions: DP845 to a Mk.XII and DP851 to a Mk.21.

December: The first production Spitfire VI (AB176) is flown.

December: By the end of 1941 the Spitfire V was in mass production and equipped no fewer than 44 RAF squadrons.

December: An order placed for 100 Spitfire VC airframes fitted with Griffon engines in its initial form with two-speed/single-stage supercharger. These emerged as the Spitfire XII.

December: Testing of Spitfire VB W3228 with a negative 'g' carburettor utilising a diaphragm system to control fuel flow is successful, the carburettor removing the previous tendency for the engine's float carburettor to cut when negative 'g' was applied. Operational advantages included providing the ability to push over or 'bunt' into a dive (something the Messerschmitt Bf 109 could do due to its fuel injected engine, a proven advantage in the Battle of Britain) and make inverted flight possible without the engine cutting. Sustained inverted flight remained impossible, however, because it also resulted in a loss of engine oil pressure, usually with expensive results.

7 December: Japan attacks the US Navy base at Pearl Harbour in Hawaii and the USA enters the war.

13 December: First flight of Mk.VB AB320, the first Spitfire equipped with the Vokes multi-vee tropical filter in a large housing under the nose. The need for such a filter arose from sand ingestion while operating in the Middle East and other dusty environments. Sand, apart from being very fine, can contain up to 85 per cent quartz, a substance harder than metal and one which caused extremely high wear in engines. A standard unfiltered Merlin was reckoned to have a life of only 36 hours in sand/dust conditions. Engine bearings and cylinder walls were the components most badly affected.

Twenty-six major modifications were required to convert a Spitfire VA or VB to tropicalised configuration including deleting the two bottom cowling panels, replacing the standard oil tank with a larger unit and replacing the air intake fairing. Trials revealed the new intake tended to work loose during dives and had to be strengthened. The Vokes filter made a substantial difference to the life of a Merlin in difficult conditions at the cost of 5-7mph (8-11km/h) in speed. AB320 was also used for trials of a 90imp gal (409 l) jettisonable overload slipper fuel tank during the first half of 1942.

A DAY IN THE LIFE

No 452 Squadron was an RAAF unit based in the United Kingdom in 1941-42 before returning to Australia to fight the Japanese. Operating as part of the Kenley Wing, by August 1941 the squadron was equipped with Spitfire VBs and featured several 'star' pilots in its number including the legendary Irishman, Brendan 'Paddy' Finucane, a flight commander with the squadron, brilliant leader of men, Battle of Britain veteran and one of the RAF's top scoring aces with 32 kills. He had attained the rank of Wing Commander by the time of his death in July 1942.

With Finucane was an impressive list of Australians, among them Keith 'Bluey' Truscott, former Australian rules football champion and the RAAF's second highest ace with 17 kills (and 452's Commanding Officer in the months before it moved to Australia), Ray 'Throttle' Thorold-Smith (8 kills) and Keith Chisholm (6) and so on.

August 1941 was a remarkable month for the squadron and took it to the top of Fighter Command's scoreboard. During that month it was credited with 22 aircraft destroyed (nine of them by Finucane) and nine probables for the loss of six Spitfires. The success continued into September (18 kills) and beyond.

The press latched onto 452's achievements, and sensing the public relations opportunities presented by this collection of Aussies and an exceptionally talented Irishman, proceeded to make Finucane and Truscott national heroes. When 'Bluey' returned to Australia in 1942 he was mobbed like a rock music star is today.

One of the squadron's pilots, Fred McCann, kept a diary of 452's activities, and his entry of 27 August understates the circumstances of his 'Circus' sortie on that day: "Got chased all over France, Dunkerque and the Channel today by hordes of 109s. Paddy saved my hide – flak terrific – Paddy got 1, Smith got 1 and I got 1...."

Years later, McCann elaborated on that sortie to Finucane's biographer, Doug Stokes, with a vivid description of what happened: "We took off in early light. Once more the bombers failed to rendezvous and we flew without them to St Omer – a popular spot! We flew into the face of the rising sun, blind spot in front and that's where the Huns had the advantage. We got a bit of flak as we crossed the French coast with some straight dotted lines of red flak among them to show the Huns where we were.

"We went on for a few minutes – just long enough to get nicely in the trap and then I saw four 109s belting down on our tail. I called a warning, then realised my R/T was unserviceable. We had been on radio silence until then, or so I thought! I got a snapshot at their last kite and he broke downwards – those 109s could roll over in a flash – then as I continued my turn tracer fairly sheeted over my cockpit.

"They had been in front of us – in the sun. I finished a 360 degree turn and had a look for the rest of our blokes – they were gone! I rolled inward and aileron turned – straightened up and had another quick look – saw a gaggle heading towards England and started to dive to bring myself under them.

"They turned towards me – 109s. I was attacked again from the sun – turn again – nothing – dive and an aileron turn from 28,000 feet down to 10,000 feet level and in a turn – four of the bastards after me, turn, dive, aileron turn. Straighten, turn, straighten – every time I turned into them they peeled away and I got a little distance ahead; a beach with a ship high and dry – now they caught me and as I turned from one, the other fired. I remember tensing as I expected to feel the shots – then I was right on the drink turning hard left and watching over my right shoulder as I outturned him.

"Suddenly he moved sideways through the air and I saw strikes all over the cockpit and rear fuselage – down he went and Paddy whizzed past followed by Smithy. I completed my turn, taking a shot at another Hun as I did so, couldn't wait to see the result and turned out to sea again. Paddy and Smithy went past again and I turned after them – more 109s in the offing – we turned out to sea and headed for home.

"I was worried about fuel. Paddy and Smithy were well ahead of me. I flew so low that the water appeared to curve away under my nose. Another attack, another break and I set out for England again, by now well behind Paddy and Smithy. The coastal guns opened up, the sea was a mass of splashes around and ahead of me and black spurts spotted the air in front.

"They hated to see me go – one more Hun attacked, red tracer floated past and he couldn't get down to me – then I was clear – throttle back and float home; over the white cliffs, they looked good, and back to Kenley.... I climbed shakily out of the cockpit..."

Spitfire VB W3821 of 452 (RAAF) Squadron on the ground at Kenley during 1941 while others from the squadron depart on an offensive sweep over France. Note the 245 Squadron Hurricane in the background. (via Mike Kerr)

The Spitfire V had to soldier on in Europe as a front line fighter for most of 1942 until the faster Mk.IX entered large scale service, the earlier model proving inferior to the Focke-Wulf Fw 190. Losses on the offensive sweeps into Europe were heavy and were stopped twice as a result. This is a Spitfire VB (W3953) of 65 Squadron with the starting trolley plugged in and the Merlin about to roar into life. (via Neil Mackenzie)

6 January: First flight of Spitfire I R6700 fitted a Merlin 60 engine, this in effect the second prototype for the Spitfire IX.

6 January: Spitfire VB AB205 flies as in effect the prototype Seafire IB with catapult spools and arrester hook fitted.

10 January: The first Spitfire carrier landing when Mk.VB BL676 with an A-frame arrester hook fitted alights on HMS *Illustrious* while it is moored on the River Clyde. Catapult launches and other deck handling trials were also conducted. The pilot was Lt Cdr H P Bramwell, commanding officer of the Royal Navy Fighter School.

22 January: The name 'Seafire' is officially adopted for the navalised Spitfire, an appropriate contraction of the conventional 'Sea Spitfire'. The name was suggested by Freda Clifton, wife of Alan Clifton, Supermarine's technical office chief.

February: With the Focke-Wulf Fw 190 appearing in ever increasing numbers, the Spitfire V was outclassed, the situation leading to the development of the Spitfire IX with 60 series Merlin engines. A Memorandum was issued during February 1942 alerting pilots to the dangers of being 'caught napping' by the superior Fw 190: "....our aircraft must fly as fast as possible in the combat zone.... [it is] essential as soon as they are liable to be detected they must open up to maximum power for formation flying.... the acceleration of the Spitfire is relatively poor. It is therefore dangerous to cruise at plus 2 boost and 1,900rpm when the Hun is about, because the time

taken to accelerate to maximum speed will allow him to draw quickly into firing range...."

As it said in *Tee Emm*: "Safety Fast – Or Prune's Guide For Living.... Don't loiter. When you can't keep up don't blame your leader; pull your finger out and cut the corners. Low revs and high boost will bring you safely back to roost. Don't wait until you see the Hun before you decide to get a move on. When you are liable to meet the enemy always fly at maximum cruising speed. If you want to live on the other side you must move fast, but equally, if you want to come back again you must save petrol."

February: Design work starts on a new wing for the Spitfire incorporating a recontoured leading edge raised by 2in (50mm). After considerable testing and

No 243 Squadron kept its Spitfire Vs (Mk.VB EN821 shown here) until June 1943. (via Neil Mackenzie)

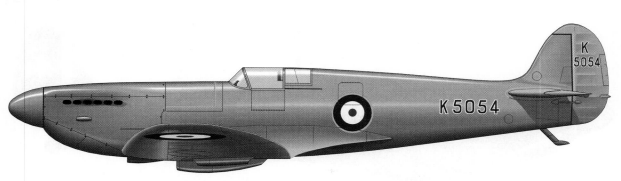

Supermarine Type 300 Spitfire prototype K5054 as configured in May 1936.

Spitfire I K9798/WZ-L of 19 Squadron RAF, Duxford UK April 1939. Note unpainted aileron, Type A1 fuselage roundel, early radio mast, flat canopy and small serial number on fin.

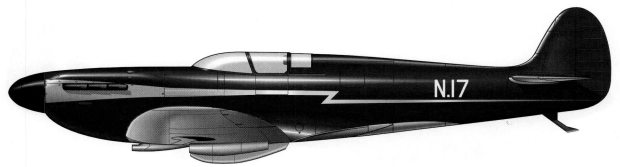

High Speed Spitfire with Class B registration N.17 (ex RAF airframe K9834) in 1939. Note tailskid and revised canopy and underwing radiator.

Spitfire IA R6800/LZ-N of 66 Squadron RAF, Gravesend UK September 1940; aircraft of Sqn Ldr Leigh.

Spitfire PR Type G (PR.VII) R7059/LY of 1 Photographic Reconnaissance Unit RAF, St Eval UK May 1941.

Spitfire IIA P7666/EB-Z of 41 Squadron RAF, Hornchurch UK November 1940; Observer Corps presentation aircraft.

Spitfire IIA P8194/DL-M of 91 Squadron RAF, Hawkinge UK April 1941.

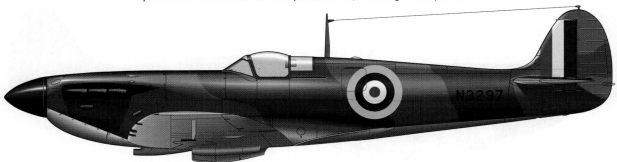

Spitfire III prototype N3297 at time of first flight in March 1940; note clipped wing and retractable tailwheel.

Spitfire F.VB AB502/IR-G of 224 Wing Desert Air Force, Tunisia 1943; aircraft of Wng Cdr Ian ('Widge') Gleed; note personalised code letters and Aboukir tropical filter.

Spitfire LF.VB AA853/WX-C of 302 (Polish) Squadron RAF, Croydon UK August 1942; aircraft of Wng Cdr Stefan Witorzenc; note temporary identification stripes for Dieppe landings.

Spitfire F.VB AB918/OU-Y 'Wellington I' of 485 (New Zealand) Squadron RAF, UK late 1941.

Spitfire F.VB '65', one of 143 handed over to the Soviet Union in March 1943; note DF loop behind radio mast.

Spitfire F.VB EP837/HL-L of 308th FS, 31st FG USAAF, Tunisia June 1943; note Vokes tropical filter.

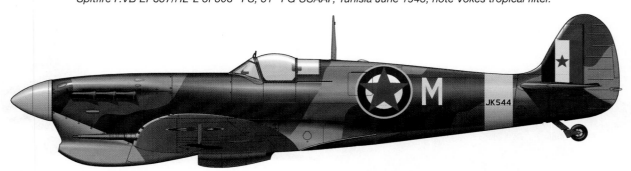

Spitfire F.VC JK544/M of 352 (Yugoslav) Squadron Balkan Air Force RAF, Yugoslavia late 1944; note Yugoslav star over RAF roundel.

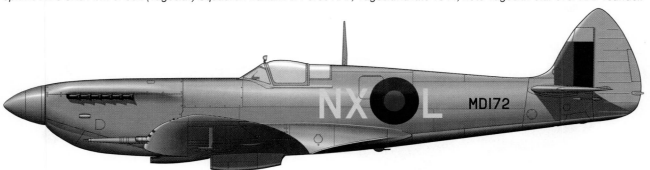

Spitfire HF.VII MD172/NX-L of 131 Squadron RAF, Culmhead UK May 1944.

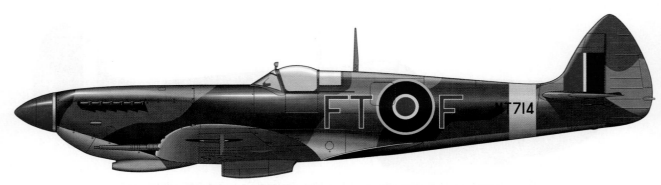

Spitfire LF.VIII MT714/FT-F of 43 Squadron RAF, France August 1944.

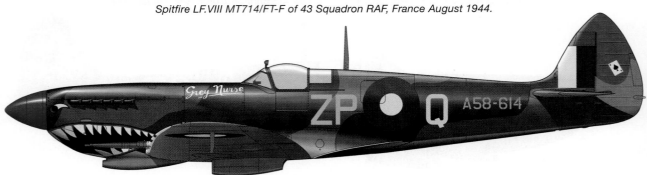

Spitfire HF.VIII A58-614/ZP-Q 'Grey Nurse' (ex MT833) of 457 Squadron RAAF, Morotai, South-West Pacific 1945.

Spitfire LF.VIII JF880/AN-U of 417 Squadron RCAF, Fano Italy December 1944.

Spitfire LF.IXE 2008 '15' of 101 Squadron Israeli Air Force, January 1949; former Czechoslovakian aircraft.

Spitfire LF.IXC MJ642/3W-II of 322 Squadron RNethAF, Twenthe 1951.

Artwork by Juanita Franzi

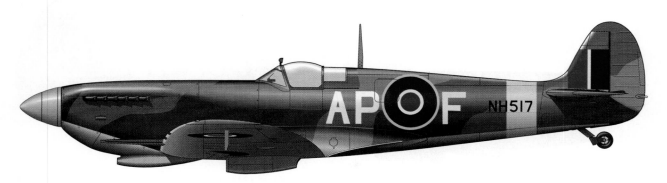

Spitfire LF.IXE NH517/AP-F of 130 Squadron RAF, Kjevik Norway September 1945.

Spitfire LF.IXC MH454/FU-N of 453 Squadron RAAF, Ford UK; note roughly applied D-Day stripes.

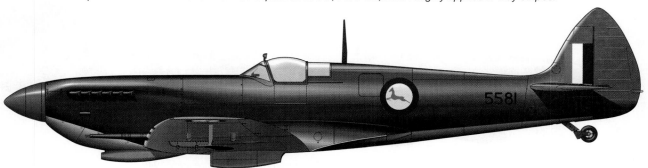

Spitfire LF.IXE 5581 of 1 Squadron SAAF, Waterkloof 1950.

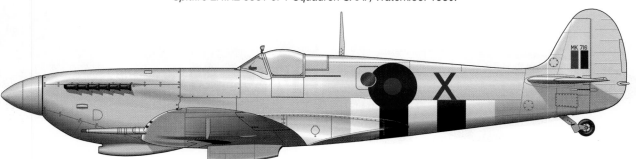

Spitfire FR.IX MK716/X of 16 Squadron RAF, Normandy September 1944.

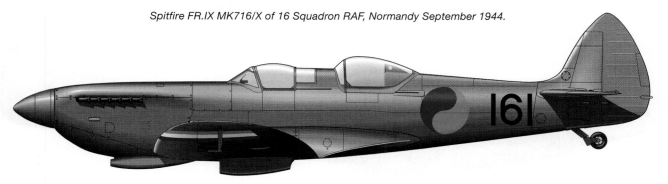

Spitfire T.9 161 (ex LF.IX PV202) of Irish Air Corps, Baldonnel June 1951.

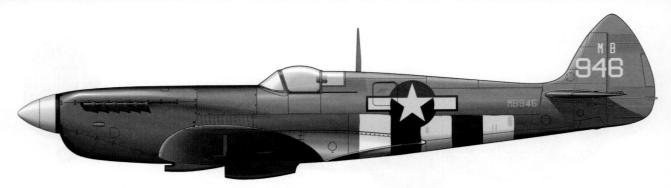

Spitfire PR.XI MB946 of 14th Photo Squadron, 8th AF USAAF, Mount Farm UK mid 1944; note deep cowling for larger engine oil tank.

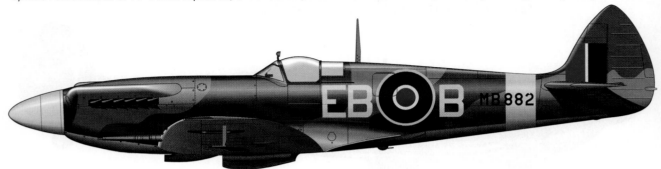

Spitfire F.XII MB882/EB-B of 41 Squadron RAF, Tangmere December 1943.

Spitfire F.XIVE RM908/UM-G of 152 Squadron RAF, Singapore early 1946.

Spitfire F.XIV RB167/DW-E of 610 (County of Chester) Squadron RAF, Friston UK mid 1944.

Spitfire FR.XIVE SM893/T of 28 Squadron RAF, Kuala Lumpur Malaysia 1946.

Artwork by Juanita Franzi

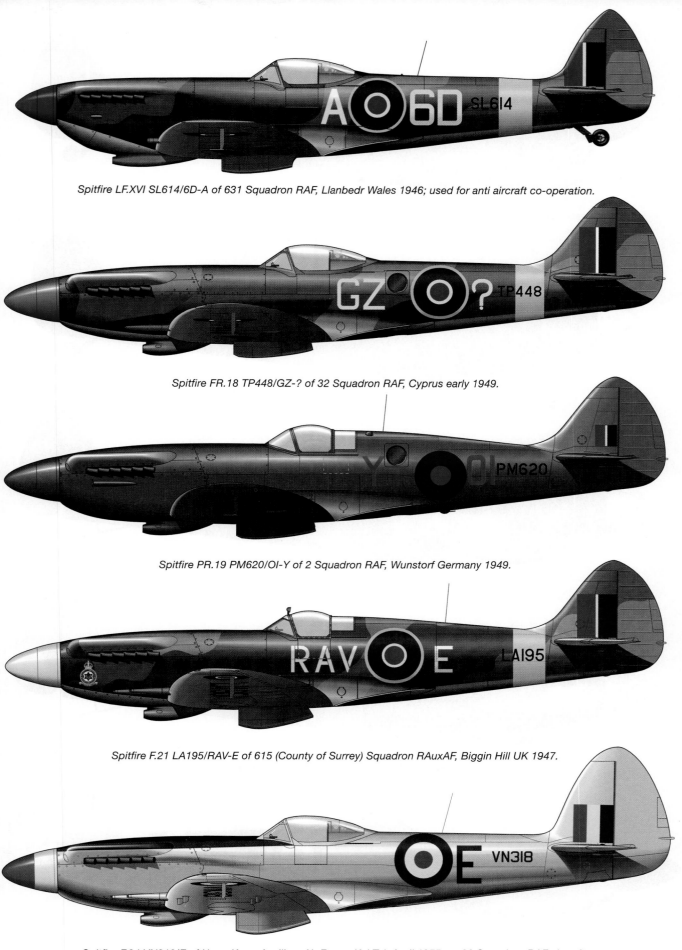

Spitfire LF.XVI SL614/6D-A of 631 Squadron RAF, Llanbedr Wales 1946; used for anti aircraft co-operation.

Spitfire FR.18 TP448/GZ-? of 32 Squadron RAF, Cyprus early 1949.

Spitfire PR.19 PM620/OI-Y of 2 Squadron RAF, Wunstorf Germany 1949.

Spitfire F.21 LA195/RAV-E of 615 (County of Surrey) Squadron RAuxAF, Biggin Hill UK 1947.

Spitfire F.24 VN318/E of Hong Kong Auxiliary Air Force, Kai Tak April 1955; ex 80 Squadron RAF aircraft.

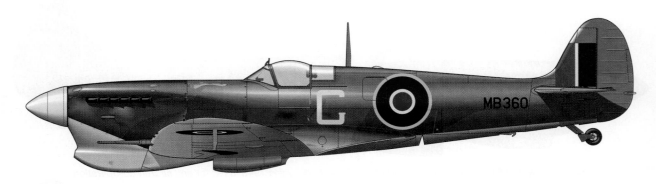

Seafire IB MB360 (ex Spitfire VB AR460) of 885 Squadron RN FAA aboard HMS Formidable *January 1943.*

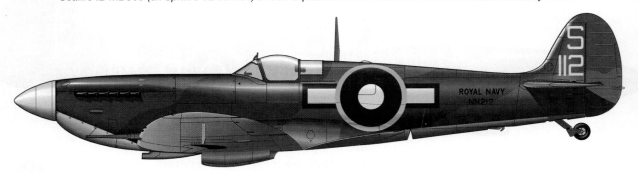

Seafire L.III NN212/S112 of 887 Squadron RN FAA aboard HMS Indefatigable *August 1945.*

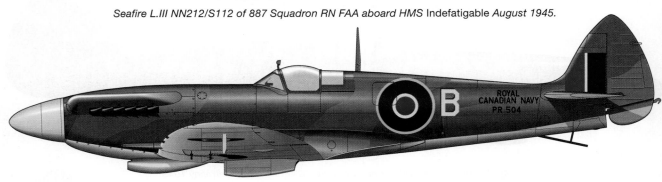

Seafire F.XV PR504/B, Royal Canadian Navy April 1947.

Seafire F.XVII SX288 of 1832 Squadron RN FAA.

Seafire FR.47 VP461 of 800 Squadron RN FAA, Korean War 1950 aboard HMS Triumph.

Artwork by Juanita Franzi

Mk.VB BL676 recorded the first Spitfire carrier landing in January 1942, alighting on HMS Illustrious and stopping thanks to the fitting of a V-frame (sometimes called A-frame) arrester hook. Like other Spitfire Vs converted to Seafire IB configuration, BL676 subsequently gained a new serial number, MB328.

revision the wing was eventually applied to the Spitfire XXI and later versions. The new wing was deemed necessary because the old one – which was later exploited to its limit on the Spitfire XIV – had a limiting speed of about 470mph (756km/h) IAS and suffered aileron control problems (including 'ballooning') at these very high velocities.

February: Spitfire VB AD371 fitted with catapult spools and arrester hook and flown as prototype Seafire IIC.

26 February: The first of two prototype Spitfire IXs converted from Mk.Vs (AB197) flies in its new guise. A second (AB196) flew on 27 March. Four other Spitfire Vs were converted to Mk.IX standards over the next few months: AA873, AB501, AB505 and AB507. Further conversion orders were placed and ultimately 5,656 Spitfire IXs were built plus 1,054 of the closely related Mk.XVI.

March: The first orders for new production Seafire IICs (based on the Spitfire VC) placed – 213 from Westland Aircraft and 202 from Supermarine. Supermarine was able to get its first aircraft off the line within two months as it did not need to tool up, as was the case with Westland which did not roll out its first example until late December 1942.

Spitfire VB/Seafire IB BL687 (later MB329) demonstrates the mid fuselage mounting point of the arrester hook.

March: The first Spitfires are deployed to the Middle East/Mediterranean Theatre – four PR.IVs (AB312, AB421, BP883 and BP904) joining No 2 Photo Reconnaissance Unit.

2 March: First production Rolls-Royce Merlin 61 for the Spitfire IX delivered.

7 March: The first operational deployment of Spitfires outside the UK when 31 tropicalised Mk.VBs (in three groups) are flown off the carriers HMS *Eagle* and *Argus* (off the Algerian coast) and flown to beleaguered Malta. A 90imp gal (409 l) belly drop tank was fitted to give the aircraft the required 660 miles (1,062km) range. Between March and October 1942, 385 Spitfire VBs and VCs had been flown off carriers for delivery to Malta, of which 367 arrived safely. Apart from *Eagle* and *Argus*, HMS *Furious* and the USS *Wasp* were also used in the operations.

23 March: After completing various trials including performing the first Spitfire landings and launches from an aircraft carrier, navalised Mk.VB BL676 emerges from Air Service Training Ltd as a fully converted Seafire IB with tropical filter, naval features and new serial number MB328. All 166 Seafire IBs were conversions of Spitfire VBs and all were allocated new serial numbers after conversion. Folding wings were not fitted to these aircraft.

27 March: The Director of Technical Development approached Supermarine on the subject of developing a folding wing naval Spitfire to solve the problem of storage on aircraft carriers, especially the smaller ones. As Supermarine's workload was substantial, the work was passed on to General Aircraft using Supermarine's design and development input. A mockup of the definitive system (upwards folding from a break just outboard of the cannon with another fold at the tips) was produced using a Spitfire I the following

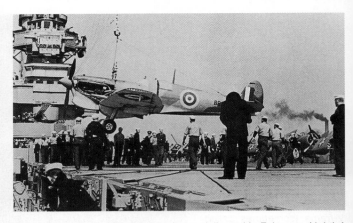

(left) Spitfire VBs undergoing final assembly at Castle Bromwich in early 1942. BL735 (second from bottom) was delivered in February. (right) A Spitfire VB bound for beleaguered Malta is loaded aboard the USS Wasp. *Note the 90imp gal (409 litres) underfuselage fuel tank, necessary for the 660 miles (1,062km) journey from the fly-off point off the coast of Algeria to Malta.*

In June 1942, Egypt based No 145 Squadron became the first Desert Air Force unit to equip with Spitfires. The aircraft are tropicalised Mk.VBs, AB326 in the foreground. (via Neil Mackenzie)

month, after which Supermarine resumed responsibility for what became the Seafire III.

31 March: First production Rolls-Royce Griffon engine delivered.

April: The high altitude Spitfire VI with lightly pressurised cabin, extended wingtips and four bladed propeller enters service with 616 Squadron based in southern England. The Mk.VI's first kill – an Fw 190 – was claimed on 18 July. Interestingly, when six were shipped to the Special Performance Flight at Aboukir (see 24 August), they could not match the performance of the 'cobbled up' high altitude Spitfire Vs prepared by that unit as they were useful only to about 35,000ft (10,670m).

18 April: Supermarine receives an order to convert 100 new Spitfire Vs to Mk.IXs in order to expedite the programme. All are delivered by the end of June with 48 of the conversions performed by Rolls-Royce.

May: No 145 Squadron becomes the first Desert Air Force unit to be equipped with Spitfires. Based at Heliopolis in Egypt, the squadron takes its tropicalised Mk.VBs into action for the first time on 1 June, providing top cover for Hurricanes on a ground strafing mission.

28 May: Maiden flight of the first production Seafire IIC built by Supermarine (MA970).

June: First low rated 1,645hp (1,267kW) Merlin 32 for the Seafire I/II/III and Spitfire PR.XIII delivered.

June: The first Spitfire IXs (converted from new Mk.Vs before delivery) are handed over to 64 Squadron at Hornchurch, this 'interim' model with a two-speed/two-stage 60 series Merlin developed quickly to counter the Fw 190's superiority over the Spitfire V. Including the very closely related Mk.XVI (which differed in having a Packard rather than Rolls-Royce Merlin), the IX went on to be the most produced of all Spitfire variants.

June: The RAAF's Nos 452 and 457 Squadrons, plus the RAF's No 54 Squadron, move from Britain to northern Australia to form the RAAF's No 1 (Churchill) Wing for the defence of the country against the Japanese.

June: Spitfire IX AB505 (one of the prototype conversions from a Mk.V) is fitted with a high rated 1,680hp (1,252kW) Merlin 77 engine to become the first HF.IX. Originally fitted with contra-props, the aircraft was refitted with the standard four bladed propeller after trials.

June: The unexpected arrival of a Focke-Wulf Fw 190 in

1942

Britain (following a navigation error by the pilot) allowed a comparative trial between it and the Spitfire V. As had been experienced in combat, the Fw 190 proved to be notably superior to the Spitfire in most respects including speed, climb, dive and roll rate performance.

15 June: The first Seafires are delivered to the Royal Navy. Interestingly, the first examples of both the converted Seafire IB (from Spitfire VBs) and the new build Seafire IIC were handed over on the same day, the result of some delays in Air Service Training's Mk.IB conversion programme and Supermarine's efficiency in building the new model. Deliveries of Spitfire IICs to Fleet

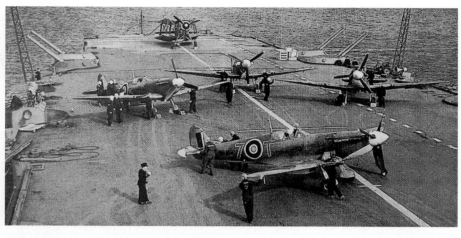

Both the converted Seafire IB and new production Seafire IIC (as here) entered Fleet Air Arm service in June 1942.

Air Arm operational squadrons began later in the same month, initially to 807 Squadron and to Nos 880, 884 and 885 by the end of August. No 801 Squadron received the Spitfire IB on its reformation in early September, the only operational FAA unit to fly this 'interim' version.

4 July: Faced with an ongoing shortage of fighters (exacerbated by slow deliveries of the interim Seafire IB and new production IIC), the Admiralty requests the transfer of 500 standard Spitfires to the Royal Navy. The Air Ministry objected, citing the need to not only supply Fighter Command but also to meet requests for Spitfires from Australia and the requirement to supply 2,400 fighters to the Soviet Union during the course of 1942. There was much heated discussion about the issue and to-ing and fro-ing on the numbers, Winston Churchill becoming involved and in effect blaming the Fleet Air Arm for its problems. Churchill had obviously forgotten that it was he, as First Sea Lord, who had cancelled the original folding wing Spitfire programme in 1940....

30 July: The Spitfire IX taken into battle (by 64 Squadron) for the first time, achieving immediate success by shooting down three Fw 190s from a formation of 15

Also entering service in June 1942 was the Merlin 61 powered Spitfire IX, its improved performance at last providing a foil for the Fw 190. Like many early examples, this F.IXC (AB196) was converted from a Mk.V and was the first conversion. It first flew in its new guise on 27 March 1942.

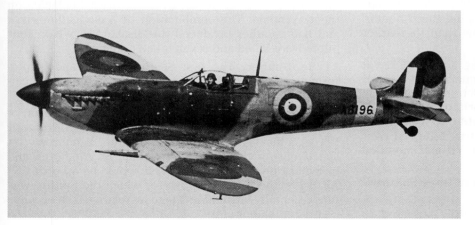

intercepted off the French coast. Flt Lt Donald Kingaby scored the first victory. It was appropriate that the Spitfire IX's combat debut should be a successful one against the Fw 190 which had hitherto been dominant over the Mk.V.

August: The first Spitfire fighter-bomber missions are performed by Mk.Vs of 126 Squadron based on Malta. Underwing racks were fitted to carry two 250lb (113kg) bombs and targets in Sicily were attacked.

August: Deliveries of Spitfire Vs to the Royal Australian Air Force at home begin. The RAAF received a total of 245 mostly tropicalised Mk.VCs and a single Mk.VB between then and November 1943. One of them (Mk.VC MH646/A58-259) was the last Spitfire V to be built at Castle Bromwich. In an unsuccessful attempt to disguise the fact that the 'world's most famous fighter' was about to arrive in Australia, the powers-that-be codenamed the aircraft the 'Capstan'. Every schoolboy and his dad knew the truth, as – presumably – did the Japanese!

August: Three Spitfire PR.IVs of No 1 PRU detached to Vaenga in northern Russia to watch German surface raiders which had been troubling convoys on their way to Russia. The aircraft flew in Soviet markings.

19 August: The greatest loss of Spitfires on a single day when 62 are lost during operations supporting the ill-fated Dieppe landings, Operation Jubilee. No fewer than 48 Spitfire squadrons were involved in the operation (42 with Mk.Vs, two with Mk.VIs and four with Mk.IXs) conducting either close air support strafing ground targets in advance of the first landings or providing fighter cover. Between them, the Spitfire squadrons flew 2,050 sorties out of the 2,600 performed by all Allied aircraft.

22 August: First flight of the true Spitfire VII prototype (AB450) with Merlin 61, pressure cabin, extended wingtips, four bladed propeller and

The Royal Australian Air Force began receiving Spitfire Vs 'at home' in August 1942. (via Neil Mackenzie)

retractable tailwheel. A standard Mk.V fuselage was used (with pressure bulkheads added), the undercarriage was from a Mk.VC and the early small fin and rudder fitted. The first production Spitfire VII (BS121) flew the following month.

24 August: First flight of the prototype Spitfire XII, the first Griffon powered version to achieve production. The prototype was converted from the first experimental Spitfire IV/XX (DP845), the type being quickly developed to counter the low level 'hit and run' raids on Britain's south coast being mounted by *Luftwaffe* Focke-Wulf Fw 190s.

24 August: A modified Spitfire VC of the Special Performance Flight in Egypt intercepts and shoots down a Junkers Ju 86P-2 reconnaissance aircraft at 42,000ft (12,800m). The Spitfire was lost during the sortie when it ran out of fuel and the pilot bailed out. The SPF was established at the Aboukir Maintenance Unit near Alexandria to intercept the high altitude reconnaissance Ju 86P-2s which were regularly overflying the Suez Canal and Alexandria. The Spitfire had its normal armament removed and replaced with two 0.50in machine guns, 'home made' extended wing tips were fitted along with a four bladed propeller and a 'souped up' Merlin 46. A second Spitfire VC was similarly modified. The Aboukir MU was also responsible for developing the 'Aboukir' tropical filter which was less bulbous than the earlier Vokes filter.

24 August: On the same day as the action in Egypt involving Ju 86P high altitude reconnaissance aircraft (see above), high flying Ju 86R bombers attack Camberley and Southampton in Britain. Spitfire Vs scrambled to intercept were unable to reach the Junkers' altitude. A similar pattern was followed over the next few weeks until the Spitfire VII and modified Mk.IXs became available (see below).

September: Maiden flight of the first of three Spitfire V floatplane conversions, W3760.

September: The first two Spitfire VIIs delivered to

Fighter Command's Special Services Flight at Northolt to counter high flying Ju 86R bombers which were operating over Britain and immune from fighters, but no interceptions were recorded. Regular squadron service doesn't start until March 1943. Two Spitfire IXs converted to HF.IX standards but with the machine guns and armour removed to save weight were also delivered to the Flight, these capable of reaching 45,000ft (13,715m). In an engagement on 12 September one of these (flown by Plt Off Prince Emanuel Galitzine) damaged a Junkers at 43,000ft (13,105m) before the Spitfire's cannon jammed. This is regardless as the highest altitude action of the war and caused the Germans to rethink the raids.

1 September: Three Spitfire PR.IVs from No 1 PRU depart Benson for North Russia where they are tasked with keeping watch on German surface raiders which were damaging convoys. The first operation took place on 10 September over Alten Fjord and the flight returned to Britain in November. One aircraft was lost when it failed to return from a sortie over Alten.

4 September: The basic unsuitability of the Spitfire for bomber escort missions due to its short range is emphasised when 11 of 12 new Mk.IXs escorting B-17s to a raid on Morlaix run out of fuel over the Brest Peninsula on the way home. The combination of a navigation error and bad weather produced the disaster, in which four pilots were killed and seven taken prisoner.

11 September: The first Seafire carrier landing performed when Capt Eric Brown alights on the escort carrier HMS *Biter*.

29 September: Nos 71, 121 and 133 'Eagle' Squadrons – manned by US pilots – become part of the US Army Air Force as the 334th, 335th and 336th Fighter Squadrons (respectively) of the 4th Fighter Group, 8th Air Force and based at Debden. Their main task would be escorting 8th AF B-17s into Europe, a task for which the Spitfire was fundamentally unsuitable due to its relative lack of range (see 4 September entry).

October: Spitfire II front line RAF squadron service ends when 504 Squadron swaps its final examples for Mk.Vs. Most squadrons had relinquished their Mk.IIs by late 1941/early 1942 but the mark continued in use throughout the war as trainers with Operational Training Units (OTUs) and the Air Fighting Development Unit (AFDU), which had some 40 on strength from early 1944 to form 'circuses' for training purposes. The Spitfire IIC (subsequently ASR.II) flew with air-sea rescue squadrons until 1944, No 277 Squadron keeping the type until May of that year.

October: The only operational Fleet Air Arm squadron fully equipped with the Seafire IB – No 801 – embarks on HMS *Furious*. No 842 Squadron had a small number on strength aboard HMS *Fencer* from mid 1943.

2 October: First flight of the 'clipped wing' Spitfire, Mk.VB AA937. The aircraft was modified by the Air Fighting Development Unit (AFDU) at Duxford without official sanction, the change involving removing the wingtips and plugging the gap with timber. Flight testing revealed a considerably improved rate of roll and greater speed at low/medium altitudes – ideal for the sort of operations many squadrons were by now flying.

Officialdom frowned on the idea at first but official testing at Boscombe Down revealed its merits. No 92 Squadron in the Middle East began clipping its Spitfire V's wingtips near the end of 1942 and on 10 April 1943 it was officially decreed that all Spitfire Vs should be so modified. No Mk.Vs were built new with clipped wings but thousands of other Spitfires (notably Mks.IX and XVI) had them either by modification in service or on the production line.

4 October: The second prototype Spitfire IV/XX (DP851) flies as the prototype Mk.21 with modified wing, beginning a lengthy period of testing and redesign which saw the Mk.21 finally enter regular squadron service in March 1945 after serious handling problems had been sorted out. In its Mk.21 guise, DP851 was fitted with extended span wings. It was written off in a landing accident in May 1943. The Spitfire 21 carried the alternative name 'Victor' for a period before the original was reverted to.

10 October: The first two RAF Spitfires arrive in the Far East – PR.IVs BP911 and BP935 – at an Indian depot for

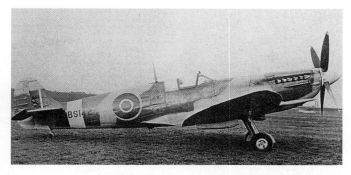

1942 was a very busy year for the development and entry to service of new Spitfire/Seafire variants. BS142 was the second production Mk.VII, delivered to the Special Services (High Altitude) Flight at Northolt in September. Regular squadron service didn't begin until March 1943.

service with No 2 PRU. The first operational sortie was conducted on 19 October over Japanese held jungle areas in Burma. It would be another 11 months before the first combat Spitfires were sent to the area.

13 October: First flight of the first production Spitfire F.XII (EN221), originally fitted with standard span wings. All but a handful of the 100 Mk.XIIs built between then and the final delivery in November 1943 had clipped wings.

25 October: First operational use of the massive 170imp gal (773 l) belly tank – which trebled the Spitfire V's standard fuel capacity – when two aircraft fly the 1,380 miles (2,220km) from Gibraltar to Malta non stop. The drag created by the tank – which wasn't jettisoned – meant the theoretical range could not be achieved, so an additional 29imp gal (132 l) tank was installed in the rear fuselage. Except for two machine guns, armament was removed to save weight and increased oil capacity was provided in enlarged lower cowlings similar to that developed for some PR Spitfires. Fifteen other Spitfires were subsequently delivered to Malta in the same manner, the flights taking about 5¼ hours. On arrival, the extra fuel and oil tanks were removed, armament restored and standard oil capacity, cowlings and tropical filters fitted.

November: Deliveries of 12 second hand Spitfire Is to Portugal begins, sufficient to equip a single squadron.

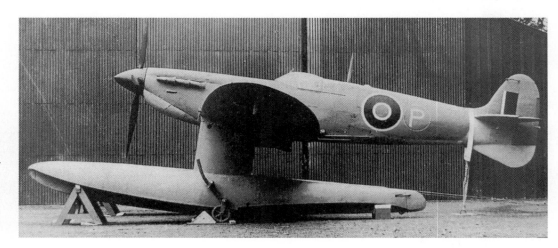

The first of three Spitfire V floatplane conversions (W3760). First flight was in September 1942. This shot shows the aircraft in its original form with standard fin and rudder. A leading edge fin extension would shortly follow. Note the four bladed propeller.

The first Griffon powered Spitfire IV (subsequently Mk.XX) was converted to serve as the prototype Mk.XII, as illustrated here. It first flew as a Mk.XII on 24 August 1942.

Intended to provide improved performance at the low altitudes at which more operations were being conducted, the first 'clipped wing' Spitfire V flew on 2 October 1942. The following year the RAF ordered that all surviving operational Mk.Vs have their wingtips removed.

(below) Fifteen Spitfire IXs were converted to PR.IX photo-reconnaissance configuration in late 1942 to help meet an urgent need for the higher performance they offered. This is EN149, first flown on 15 December 1942.

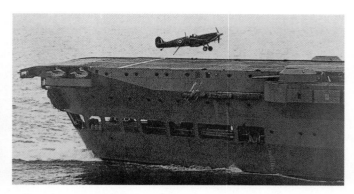

(left) The Griffon powered Spitfire XII began to roll off the production line in October 1942. EN223 was the third production example. (right) The Seafire's first taste of combat was in November 1942 during Operation Torch, *the Allied invasion of North Africa.*

November: First operational use of the photo-reconnaissance Spitfire PR.IX (15 converted from Mk.IX fighters) by No 541 Squadron based at Benson.

8 November: The start of Operation *Torch*, the Allied invasion of North Africa through landings in Algeria and French Morocco and the first action in which the Seafire was involved. Early RN FAA Seafire participating squadrons were Nos 880 (aboard HMS *Argus*), 801 and 807 (HMS *Furious*), 884 (HMS *Victorious*) and 885 (HMS *Formidable*). Sub Lt G C Baldwin of 807 Squadron recorded the Seafire's first confirmed 'kill' on 8 November, a Vichy French Dewoitine D.520 fighter. Sub Lts A S Long and J D Buchanan from 885 Squadron had earlier in the day scored a 'probable' Martin 167 bomber.

9 November: Testing of a Seafire with folding wings begins, the prototype converted from Seafire IIC MA970, the first production aircraft. In production form, the folding wing Seafire was designated the Mk.III powered by a 1,470hp (1,096kW) Merlin 55 driving a four bladed propeller.

20 November: Maiden flight of the first Spitfire VIII (JF724).

21 November: Maiden flight of the first Spitfire PR.XI (BS492).

December: AVM Sir Keith Park, Commander-in-Chief Malta, writes to the Air Ministry urging the production of conversion kits for the Spitfire allowing it to carry two 250lb (113kg) bombs under the wings. A locally developed kit was fitted to some of Park's Spitfires, the C-in-C noting that 50 per cent of his aircraft used on fighter sweeps over enemy airfields south of Sicily had them fitted. The bombing technique was to perform a stall turn over the target, dive at an angle of about 74 degrees and release the bombs at between 7,000 and 12,000 feet (2,130-3,660m) depending on the amount of anti-aircraft fire. An additional 500lb (227kg) bomb under the fuselage was also tested, giving the 'Spit-bomber' a total load of 1,000lb (454kg).

December: The low rated 1,645hp (1,227kW) Merlin 32 engine tested in Spitfire I L1004. The enhanced low altitude speed it produced and increased rate of climb made it ideal for the Seafire, Mk.IICs thus equipped designated L.IIC by the Royal Navy.

December: The first Spitfire IXs are shipped to the North African theatre to help counter the recently arrived Focke-Wulf 190 which was repeating the impact it had made in Europe. The first Mk.IXs were attached to 145 Squadron, No 81 Squadron took them into action for the first time in January 1943.

THE PLASTIC SPITFIRE

Fears in 1940 that supplies of sheet aluminium could be threatened due to the loss of French controlled bauxite deposits led to investigations into building Spitfires (and other aircraft) from plastics. After examining several alternatives, it was decided to build two Spitfire fuselages from Gordon Aerolite, a plastic consisting of fibres of flax impregnated with phenol resin.

The fuselages were built in 1941-42 and found to be of similar weight and strength to the standard item. Although time consuming and costly, the experiments provided much useful information which would be later applied, especially in the area of synthetic glues. By the time the fuselages were built the supply of raw materials for the production of sheet aluminium was secured so there was never any real intention to mass produce 'Plastic Spitfires'.

The first Spitfire PR.XI flew in November 1942.

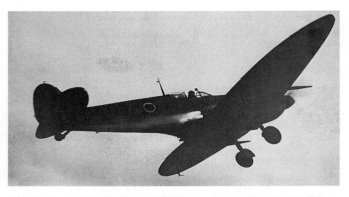

The use of Rocket Assisted Takeoff Gear (RATOG) was tested on a Seafire IIC in February 1943, the aim to make operations from smaller aircraft carriers more feasible. Although successful, the system was not used operationally during the war, largely due to deck safety considerations.

January: Initial orders for the Seafire III placed with Westland (200) and Cunliffe-Owen Aircraft (250).

29 January: Following requests from the Admiralty, Supermarine releases a specification for a navalised version of the Griffon powered Spitfire XII with folding 'C' wing as used on the Seafire III. Powerplant was a 1,815hp (1,353kW) Griffon VI with single-stage supercharger driving a four bladed propeller, the fuselage was also similar to the Seafire III and features of other variants (including the Spitfire VIII's tail surfaces but with broad fin and rudder, retractable tailwheel and the Mk.IX's wing root fuel tanks) were incorporated.

February: A Seafire IIC fitted with rocket assisted takeoff gear (RATOG) is trialed aboard HMS *Illustrious*. The tests were regarded as successful but although subsequent production Seafires were fitted with RATOG attachment points, the method was never used operationally during the war.

Intended to counter the low level 'hit and run' raids being carried out by Focke-Wulf Fw 190s over southern England, the Spitfire XII entered squadron service in February 1943 as the first operational Griffon powered Spitfire.

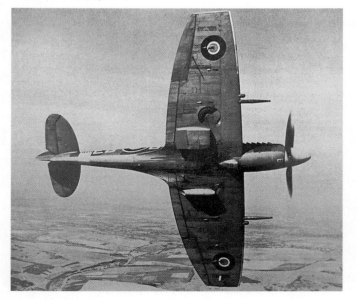

6 February: The Spitfire's first 'kill' in the Pacific theatre when a Mk.VC of 54 Squadron RAF (based in Darwin, Australia) flown by Flt Lt Foster claims a Mitsubishi Ki.46 'Dinah'. No 54 Squadron flew as part of No 1 (Churchill) Wing along with the RAAF's Nos 452 and 457 Squadrons, all equipped with Spitfire VCs. The wing was commanded by Wng Cdr Clive Caldwell, Australia's top scoring fighter ace with 28½ victories to his credit by the end of hostilities.

February: The Griffon powered Spitfire XII enters service with No 41 Squadron.

March: Start of the conversion programme re-engining most Merlin 46 powered Seafire I/IIs with the much more effective low rated Merlin 32.

March: 143 Spitfire VBs handed over to the Russian Mission at Basrah (Iraq) for use by the Soviet Air Force to help meet its apparently insatiable appetite for fighters. They are the first of 1,331 Spitfires delivered to the Soviet Union between then and April 1945, the remainder comprising 1,186 LF.IXs and two HF.IXs from mid 1944.

The first of an eventual 1,331 Spitfires for the Soviet Union was handed over in March 1943.

March: The first Castle Bromwich built production Spitfire IX (JK395) rolls off the Birmingham line. Another 5,094 would follow between then and 1945.

March: Three prototype Griffon powered Seafire XV prototypes ordered from Supermarine with another three ordered in May. Production aircraft would be built by Westland and Cunliffe-Owen.

March: The Spitfire VII enters service with 124 Squadron at North Weald.

9 March: Eric Brown performs 15 deck landings with the Spitfire XII on HMS *Indomitable*, part of the preparations for the eventual introduction of the Griffon powered Seafire into Fleet Air Arm service.

April: In conjunction with the clipped wings developed for the Spitfire to improve its roll rate at the expense of altitude performance, tests begin of the Merlin 50M with a smaller diameter supercharger impeller which provided maximum boost (+18lb) and power at low altitude. Equipped with the negative 'g' carburettor, the 50M produced 1,585hp (1,182kW) at only 3,800ft (1,160m). The Merlin 45M (with standard carburettor) and 55M (two piece cylinder block) were similarly modified and rated, these the equivalent of the standard Merlin 45, 50 and 55 with normal impeller. Spitfire LF.Vs modified with clipped wings and 'M' series engines were referred to as being 'clipped, cropped and clapped', the latter alluding to their often well used condition when they were converted.

18 April: The 'Palm Sunday Massacre' in North Africa when 47 USAAF Curtiss P-40 Warhawks and 12 Spitfires from the RAF's 92 Squadron intercept a formation of 90 Junkers Ju 52/3m transports escorted by 50 German and Italian fighters. The final score was 77 enemy aircraft destroyed for the loss of six P-40s and one Spitfire. Four days later, RAF Spitfires and South African Air Force Kittyhawks shot down 14 Me 323 heavy transports and seven of their fighter escort in a single action off Cape Bon. The Me 323s were all loaded with fuel and the sea into which they crashed was ablaze for many hours afterwards. These actions were part of a concerted effort to destroy the *Luftwaffe* in North Africa and to hasten the demise of the *Afrika Corps* by starving it of supply.

23 April: The first Westland built Seafire III (LR765) leaves the production line. Some early aircraft featured non folding wings and were initially designated Mk.III (Hybrid) and then Mk.IIC (Hybrid).

May: A Spitfire PR.IX of No 542 Squadron (flown by Flg Off F D Fray) brings back the famous photographs of the breached Moehne and Eder Dams after 617 Squadron's 'Dambusters' attack.

13 May: The second Griffon powered prototype (DP851 – by now operating as the Mk.21 prototype) – written off after its undercarriage failed while landing.

15 May: The high altitude Spitfire VII achieves it first kill when a 124 Squadron aircraft destroys an Fw 190 at 38,000ft (11,580m) over Plymouth. The squadron was the first to equip with the Mk.VII, in the same month.

18 May: Supermarine issues its Specification 472 for the Type 379 Spitfire XIV, basically a Spitfire VIII powered by a 2,050hp (1,529kW) Griffon 65 engine with two-speed/two-stage supercharger driving a five bladed Rotol propeller. The Griffon 75 for low altitude operations was also mentioned (this was never introduced) as was the 'C' wing armament of two 20mm cannon and four 0.303in machine guns; the possible need to introduce larger vertical

SPITFIRE XIV PROTOTYPES

With the Rolls-Royce Griffon in production by the end of 1942, it was decided to produce six testbed Spitfires (JF316-JF321) powered by this engine as part of the development programme for models which were more advanced than the initial Mk.XII which was by then about to enter limited service. These six aircraft were in effect the prototypes for what turned out to be the major production Griffon powered Spitfire, the Mk.XIV.

JF316: First flight 20 January 1943 with 1,735hp (1,294kW) Griffon IV and four bladed propeller (as on Mk.XII), subsequently Griffon 61 and five bladed propeller as Mk.XIV.

JF317: First flight 4 March 1943; Griffon 61 and five bladed propeller; Mk.VIII wing; extended cowling; larger air intake; larger fin and rudder; later fitted with Griffon 65; also used to test contra-props.

JF318: First flight 29 May 1943; Griffon 61; later fitted with extended wing tips.

JF319: June 1943; Griffon 65; universal wing; enlarged fin and rudder intended for F.23 Valiant (the proposed laminar flow wing Spitfire).

JF320: June 1943; Griffon 61; later tested contra-props.

JF321: First flight 20 August 1943; Griffon 61; later tested contra-props.

JF318, the first of six Spitfire VIIIs converted to Griffon power to act as prototypes for the Mk.XIV.

A Spitfire VB with the compact tropical filter developed by the RAF Maintenance Unit at Aboukir in Egypt. The aircraft is AR287.

The first Seafire III left Westland's production line in April 1943. This was the first Seafire with folding wings, the geometry demonstrated here by Seafire IIC MA970 after conversion to the new model.

tail surfaces (subsequently applied); the fact that some aircraft may be fitted with contra-props; and the planned introduction of the cut down rear fuselage decking and bubble canopy (also subsequently applied).

23 May: Spitfire VC MA648 becomes the first Spitfire LF.IX with the fitting by Rolls-Royce of a low rated Merlin 66. The LF became the most numerous of the Spitfire IX variants, accounting for 4,010 aircraft or 71 per cent of Mk.IX production.

June: Production of the Spitfire IX at Supermarine's factories ends after 561 are built – all Mk.IX production now comes from Castle Bromwich.

The Spitfire IX entered production at Castle Bromwich in 1943, this shot showing more than 40 of them awaiting collection. Three Avro Lancasters are also visible, the factory building 300 of the bombers along with 11,989 Spitfires and Seafires.

June: Formation of the 2nd Tactical Air Force, the RAF organisation which would provide the air element supporting the planned invasion of Europe.

July: Seafire IIC MB194 reflown as prototype LR.IIC with one vertical and one oblique F.24 camera with armament retained. This and subsequent conversions were performed by Heston Aircraft.

11 July: Two days after the start of the invasion of Sicily, a 72 Squadron aircraft flown by Flg Off D N Keith becomes the first Spitfire to land on the island. After destroying two enemy aircraft in combat, Keith had run out of fuel and put down on the ploughed up airfield at Pachino which was still being prepared by Royal Engineers. The airfield was serviceable two days later and quickly occupied by numerous RAF and USAAF fighter squadrons.

24 July: First flight of the production prototype Spitfire 21 (PP139) with definitive broad fin and rudder, four cannon armament and modified wing, albeit with extended tips which were subsequently deleted. Orders for some 3,000 Spitfire 21s were placed in 1943-44 but only 120 production aircraft were ultimately built.

August: Britain and neutral Portugal strike a pact which grants Britain access to aircraft and shipping facilities in the Azores in exchange for protection should Portugal be attacked by Spain or Germany and the supply of military aid. Included in the deal are an initial 33 ex RAF Spitfire VBs, deliveries of which begin in October. Others are supplied to Portugal later.

August: The Spitfire VIII enters operational service with No 145 Squadron based in Italy. Although production aircraft had begun rolling off the line in late 1942/early 1943, the initial rate was slow and the fact that the type was almost exclusively sent to overseas based squadrons meant that it took a few months for the first unit to become operational.

September: The first combat Spitfires arrive to fight in the India-Burma campaign, Mk.VCs replacing the Hurricanes of Nos 136, 607 and 615 Squadrons. The latter pair are deployed to Chittagong where they quickly destroy four Japanese Mitsubishi Ki-46 'Dinah' reconnaissance aircraft. The very first Spitfires deployed to the Far East theatre were two PR.IVs in October 1942.

The Spitfire VII entered regular squadron service in March 1943. F.VII MD159 is pictured here with a Handley Page Halifax bomber, the photograph possibly taken while the Spitfire was serving with No 1420 Meteorological Flight late in its career. (via Neil Mackenzie)

Like the previous year, 1943 was also very busy as far as the introduction and development of new Spitfire variants is concerned. One new model to appear (first flight May 1943, converted from a Mk.VC) was the mass produced LF.IX with a low altitude rated Merlin 66 engine. This is a production LF.IX (MK126), built right at the end of the year and delivered in January 1944. (via Neil Mackenzie)

Seafire operations began in earnest during 1943, these two shots illustrating the aircraft on and above HMS Indomitable. Note the rudder deflection on the aircraft taking off as the pilot counters the engine's torque.

September: Spitfire VIII JF299 is tested as the first Spitfire fitted with the cut down rear fuselage decking and teardrop canopy which would subsequently be fitted to late production aircraft of several models.

9 September: Start of the Allied landings at Salerno with 106 Seafires from the carriers *Attacker*, *Battler*, *Hunter*, *Stalker* and *Unicorn* conducting 265 sorties on the first day. After only two days, no fewer than 67 had been lost, 42 of them due to landing accidents on the carriers (see *Sorting The Seafire* box story).

October: Deliveries of Spitfire VIIIs to the Royal Australian Air Force begin, a total of 410 supplied between then and June 1945. About 100 of the later arrivals were delivered direct to storage and few flew operationally before being scrapped. They were used mainly on the 'island hopping' operations to the north of Australia as the Japanese were gradually forced back but were badly underemployed, the result of General Douglas MacArthur's policy to exclude non Americans from his obsessive 'I will return' to the Philippines.

October: Plans for a floatplane fighter based on the Spitfire VIII mooted but after investigation, not proceeded with.

The captured Spitfire VB (EN830) re-engined with a Daimler-Benz DB 605A engine and evaluated by the engine manufacturer's flight test department in November 1943.

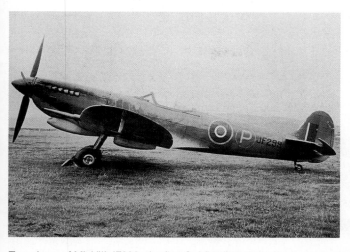

Two views of Mk.VIII JF299, the first Spitfire fitted with the 'rear view' fuselage and bubble canopy. It first flew in the new configuration in September 1943.

October: Production Spitfire XIVs begin rolling off the line, the first aircraft RB140.

October: The first Cunliffe-Owen built Seafire III (NN333) completed.

30 October: The three Spitfire V floatplane conversions (W3760, EP751 and EP754) arrive at Alexandria, Egypt, the intention being to operate them against *Luftwaffe* Ju 52 transports flying from Greece to Crete. None saw operational service and they were returned to Britain.

November: Spitfire IX JK535 is flown with six bladed contra-props as part of the programme to nullify the effects of torque. JL349 followed in March 1944.

November: Operation *Pointblank* begins, as part of the 'softening up' process in preparation for the invasion of Europe. The operation was designed to reduce the number of *Luftwaffe* fighters available to defend the invasion. By March 1944 2,950 had been destroyed.

NS487, the prototype Seafire XV with Griffon engine. It was rolled out in November 1943.

SORTING THE SEAFIRE

One of the problems which marred Seafire operations during much of its wartime service was the very high incidence of deck landing accidents, often as a result of the original vee-frame arrester hooks pulling out of their mountings or pilots missing the wires and ploughing into their carriers' crash barriers.

The problem came into focus in early September 1943 during Operation *Avalanche* – the landing of Allied forces at Salerno – when the original strength of 106 Seafires from five escort carriers was reduced to only 39 two days after the operation began. No fewer than 42 Seafires were written off due to deck landing mishaps.

In September 1943, Eric Brown began a series of trials using a Seafire L.IIC designed to come up with ideas to reduce the problem. Jeffrey Quill completed them, their recommendations including strengthening the tail hook assembly, teaching pilots to use a curved rather than crabbed straight in approach, fitting multi ejector type exhaust manifolds and training pilots to land with their heads out of the cockpit, looking along the port side of the engine cowling.

Quill also noted the Seafire's poor inherent stability at low speeds and the fact that generally, it lacked the robustness necessary for carrier landings. It was suggested that a 'sting' type arrester hook would help reduce the accident rate, but this did not appear until the Griffon powered Seafires came along.

The introduction of the more completely 'navalised' Merlin powered Seafire III in 1943 did little to reduce the problem, despite the growing experience of pilots. The problem centred around the fact that the Seafire in its early forms was basically a simple adaptation of the land based Spitfire, and as such was being asked to operate under extremely demanding circumstances for which it had not been designed. The Griffon powered Seafire XVII was the first variant to be fully modified for naval use including the fitting of a much stronger and long stroke undercarriage. It was, however, a post war aircraft.

A not untypical scene during early Seafire operations.

Some open air maintenance for the Spitfire VCs of No 253 Squadron in Italy towards the end of 1943. EF553 in the foreground is a Westland built example. (via Neil Mackenzie)

Another new Spitfire model to enter squadron service in 1943 was the Mk.VIII in August. JF880 is an LF.VIII delivered the following month.

1 November: The proposed name 'Victor' is officially dropped for the new generation of Spitfire developments, the first of them becoming the Spitfire F Mk.21.

15 November: Inexplicably, unpopularly (and temporarily), RAF Fighter Command is renamed Air Defence Great Britain (ADGB). At the same time, a large number of its squadrons are transferred to the newly created 2nd Tactical Air Force, part of the organisational structure for the planned invasion of Europe. Half of the 2nd TAF's fighter squadrons were equipped with either Spitfires or Seafires. These were Nos 4, 16, 26, 56, 63, 66, 91, 124, 132, 222, 302, 308, 310, 312, 313, 317, 322, 329, 331, 332, 340, 341, 349, 400, 401, 403, 411, 416, 431, 442, 443, 453, 485 and 602 (total 34 squadrons) with Spitfires plus the RN Fleet Air Arm's Nos 808, 885, 886 and 897 with Seafires. ADGB squadrons at the time equipped with Spitfires were Nos 1, 33, 41, 64, 74, 80, 126, 127, 130, 165, 229, 234, 274, 303, 345, 350, 402, 501, 504, 610, 611 and 616.

November: The first Seafire XV (NS487) completed.

November: All Allied forces in the China-India-Burma theatre come under the control of the newly established South-East Asia Command (SEAC).

18 November: A captured Spitfire VB (EN830) is delivered to the Daimler-Benz flight test department at Echterdingen near Stuttgart where a 1,475hp (1,100kW) DB605A inverted vee-12 engine from a Messerschmitt Bf 109G is fitted in place of the original Merlin 45 for comparative testing. The conversion was relatively simple and only took about a fortnight to perform, one of the major tasks being to replace the Spitfire's original 12 volt electrical system with a *Luftwaffe* 24 volt system. A VDM propeller from a Bf 109G was also fitted. Flight testing revealed the Daimler-Benz Spitfire to be slightly faster than the original at low altitudes and about the same higher up, but most of the test pilots' praise centred around the handling of the Spitfire compared to the 109.

Willy Ellenrieder was the first to fly the aircraft (before and after conversion), reporting: "Well, this Spitfire was a dream, a pilot's aircraft. It was my real baby. I had never before flown an aircraft like this. We were accustomed to the Bf 109's ill manners; bad visibility before takeoff because the nose pointed skywards, the

THE AMERICAN MERLIN

Rolls-Royce and Packard concluded their agreement for the manufacture of Merlins in the USA in July 1940. The British manufacturer had originally wanted Ford to build the engine at its French subsidiary and after the fall of France in June 1940, in the USA. Henry Ford thought that Britain would be defeated by Germany and decided not to participate in the Merlin programme.

The initial contract with Packard covered 1,500 engines with options on a further 10,000. Packard ended up building 55,523 Merlins at an average price of $US12,000 each, the figure representing almost exactly one-third of all Merlin production.

The first Packard-Merlin V-1650-1 was bench tested in August 1941, subsequent production increasing steadily to 600 per month by July 1942, 1,300 per month by July 1943 and peaking at 2,700 per month in July 1944. Total production for that year reached 23,169 engines at an average of 1,930 per month. The rate declined after that, reducing to 1,500 per month in July 1945.

Packard introduced several important innovations to the Merlin including the Bendix-Stromberg injection carburettor, automatic supercharger gear shifting, water-alcohol injection, ball bearing main water pump (replacing Rolls-Royce's plain bearing unit), centrifugal air/oil separator to prevent foaming, and a new high altitude magneto.

frightening takeoff with the common tendency to swing, the unwieldy handling during the final approach and the bumpy landings...."

December: Spitfire LF.IX MJ556 is completed as the first Spitfire XVI with Packard-Merlin 266 engine.

December: The first RAF squadrons equipped with Spitfire VIIIs (Nos 81 and 152) arrive in India, transferred from Italy.

10 December: Allied air units in the Mediterranean area come under the control of a new organisation, the Mediterranean Allied Air Forces.

PP139, the production prototype for the Spitfire 21 and first flown on 24 July 1943. Fitted with a new wing, development was troubled and the F.21 didn't enter service until 1945.

The Spitfire XIV entered service with 610 Squadron in January 1944.

January: The Seafire begins operations in the Far East, flying from the carriers *Battler* and *Atheling* on 'trade protection' duties against German and Japanese submarines which were attacking convoys in the Bay of Bengal area. Five carriers were involved in all and two squadrons were equipped with Seafires – No 834 with a mixed fleet on *Battler* and No 889 aboard *Atheling*.

January: Spitfire XIV RB141 is tested by the Air Fighting Development Unit (AFDU) at Duxford, comparing it to the Spitfire IX and a captured Messerschmitt Bf 109G: "The Spitfire XIV is superior to the Spitfire IX in all respects. It has the best all-round performance of any present day fighter, apart from range". Compared to the 109G: "The Spitfire XIV is 40mph (64km/h) faster at all heights except near 16,000ft (4880m) where it is only 10mph (16km/h) faster. Climbing at full throttle it draws away from the 109G quite easily. The Spitfire XIV easily out-turns the 109G in either direction; it is superior in every respect."

2 January: The Spitfire XIV enters service with 610 Squadron at Exeter.

27 January: Maiden flight of the first production Spitfire 21 (LA187), initially completed to an interim standard with Mk.VC wings and a smaller fin and rudder. The

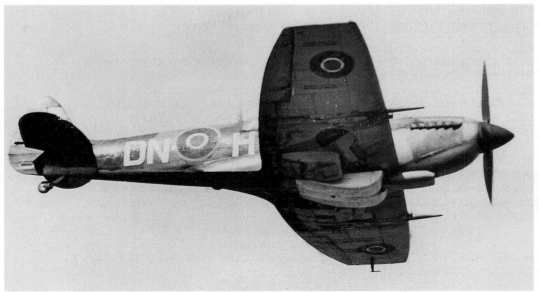

Clipped wings and low rated engines were de rigeur for Spitfires from 1944 as the majority of missions were flown at low altitudes. This is an LF.IXE of 416 (Canadian) Squadron operating in the European Theatre. The 'E' wing retained the two 20mm cannon but the four 0.303in machine guns were replaced by two heavy calibre 0.50in weapons. (via Neil Mackenzie)

Seafire operations in the Far East began in early 1944. This group of photographs shows some of the shipboard activity as aircraft are prepared for launch. (via Neil Mackenzie)

first five aircraft (LA187-191) were extensively used for trials, as were others. Several were fitted with Rotol six bladed contra-props, this at one stage intended as the standard fitting.

March: Seafire III NF545 flies as the prototype LF.III with low rated 1,585hp (1,182kW) Merlin 55M.

4 April: The first of 16 photo-reconnaissance Spitfire PR.Xs (MD192) with pressurised cabin delivered, the remainder handed over between then and mid May.

15 April: The Fleet Air Arm's No 899 Squadron begins carrier operations (training aboard HMS *Khedive*) with

the Seafire III, the first fitted with folding wings. The FAA had been long crying out for a folding wing naval Spitfire (the Seafire I and II had fixed wings) and could have had it three years earlier if the original proposals had not been cancelled by Winston Churchill (then the First Sea Lord) in early 1940.

May: First Spitfire PR.XIX delivered to 542 Squadron at Benson. The first 22 aircraft had unpressurised cockpits.

May: Spitfire LF.IXB MJ892 converted to a floatplane fighter. Trials were completed but no production undertaken.

A MATTER OF BOOST

The use of the word 'boost' when talking about engine superchargers such as those installed in the Rolls-Royce Merlin and Griffon results from the supercharger impeller which was driven off the rear of the engine's crankshaft through a series of gears. The impeller established a positive pressure in the engine's induction system.

This positive pressure was referred to as 'boost' and measured in pounds per square inch (psi) by the British and in inches of mercury by the Americans.

The standard atmospheric pressure of around 15psi was taken as the datum point and called zero by the British, variations up and down measured as plus or minus so many pounds of boost. When an engine is noted as producing 1,546hp (1,167kW) at +15lb boost, for example, it is really operating at +30lb taking the standard atmospheric pressure into account.

Similarly, cruise power settings may see a negative boost figure. The minimum fuel consumption setting for the Spitfire V with Merlin 45 engine is listed as -4lb boost at 1,800rpm, the figure implying an impossible negative boost figure but in reality meaning +11lb.... all perfectly simple!

The Americans take 30 inches of mercury as the datum point or zero (standard atmospheric pressure again) with each pound of boost equal to about 2 inches of mercury. Therefore a Rolls-Royce Merlin 66 produced its maximum power at +18lb boost while the US built Packard Merlin 266 equivalent produced the same power at 66 inches of mercury, remembering that 30in is the starting point, 18 x 2 = 36, and 36 + 30 = 66.

The Germans measured their engine boost pressures in 'atmospheres' but convention – and the need to avoid further complication – has ensured that Merlin and Griffon boost pressures are expressed in pounds or inches and not their metric equivalents.

The sole Spitfire IX floatplane conversion (MJ892) flew in May 1944.

May: The 140th and last Spitfire VII (MD190) delivered.

June: The third prototype Seafire XV (NS493) modified as the first Seafire XVII by Westland with bubble canopy and cut down rear fuselage decking, strengthened long stroke undercarriage and rear fuselage fuel tank. Otherwise similar to the Mk.XV, production Seafire XVIIs also had a 'sting' type arrester hook.

4 June: Two days before D-Day and all the aircraft involved with the Allied Expeditionary Air Force and the invasion of Europe are painted with conspicuous black and white stripes on their fuselages and wings. By this time the RAF's 2nd Tactical Air Force had 34 RAF Spitfire squadrons and four Fleet Air Arm Seafire squadrons on strength, while the Air Defence of Great Britain organisation had 22 Spitfire squadrons.

6 June: D-Day and the invasion of Europe – Operation *Overlord* – begins. The claim for the first Allied aircraft to 'land' in France probably goes to Spitfire IX MJ339 of 602 Squadron. Flown by the Canadian Plt Off D L Kidd, the aircraft suffered an engine failure over the bridgehead area and Kidd was forced to make a successful 'dead stick' belly landing among a group of American

The first production Spitfire 21 (LA187) flew in January 1944, completed to an interim standard.

1944

All aircraft involved with the Allied Expeditionary Force had these 'D-Day stripes' painted on two days before the big event to help identification. This is a PR.XI: the camera ports can be seen under the rear fuselage.

soldiers who were under heavy fire at the time. Kidd sent a message to his squadron saying he was safe and then quickly left his Spitfire to find cover.

D-Day saw nine Spitfire squadrons provide top cover for the landing troops while 23 Spitfires from Nos 441, 442 and 443 (Canadian) Squadrons attacked the Wurzburg radar installations at Cape D'Antifer with 500lb (227kg) bombs on the same day.

10 June: The first RAF squadrons land in France after the D-Day landings, Nos 130 and 303 putting their Spitfire Vs down on a strip near the Gold Beach area.

MORE THAN ENOUGH

The ongoing quest for more performance from the Spitfire relied heavy on its Rolls-Royce Merlin (and Griffon) engine producing more power. The introduction of the Merlin 60 series with two-speed/two-stage supercharger made a significant contribution and the maximum boost pressures at which the engines could operate increased as time went on.

The original 1,030hp (768kW) Merlin II and III in the Spitfire I ran at a maximum boost of +12lb on 100 octane fuel (87 octane and +6lb was used at first); the Merlin 63/66 models as used in the Spitfire VIII and IX turned out up to 1,710hp (1,275kW) on +18lb boost.

'Basta' was the codename used for Merlin engines modified to permit operation at +25lb using 150 octane fuel. Apart from being a way to improve the Spitfire's performance generally, it was subsequently applied to some aircraft in order to increase their speed at low level so they could catch V-1 flying bombs. In a Merlin 66 this resulted in a maximum power output of over 2,000hp (1,491kW) with commensurate performance increase, although the cost to these especially modified engines was a limited service life.

The world 'Basta' is Italian for 'more than enough' and co-incidentally was also Germany's codename for Gibraltar.

The second Spitfire VIII (JF275) had its original Merlin 61 replaced with a Basta low rated Merlin 66 and tested. The result was an increase in maximum speed from 338mph (544km/h) to 362mph (582km/h) at sea level and from 380mph (611km/h) to 409mph (658km/h) at 14,000ft (4,270m). Maximum rate of climb increased from an already impressive 4,610ft (1,405m)/min to a quite remarkable 5,580ft (1,700m)/min. Other testing saw a Basta equipped Mk.VIII record a level maximum speed of 445mph (716km/h) at 25,000ft (7,620m).

Deliveries of the Packard Merlin powered Spitfire LF.XVI began from Castle Bromwich in October 1944. LF.XVIE SL576 (built in mid 1945) was typical of most of this final Merlin powered variant with clipped wings, bubble canopy and pointed rudder.

Sovereign 1 – Spitfire   109

LATE OPERATIONAL ENHANCEMENTS

Apart from the major engine, airframe and performance developments which were applied to the Spitfire during its production life, the aircraft was also subject to numerous upgrades of other items of equipment and its specification in order to make it an even more effective weapon.

Four of the significant changes which were incorporated in the latter stages of the war were the gyroscopic gunsight, the bubble or teardrop canopy, increased internal fuel capacity and the 'E' wing with its revised armament.

Gyroscopic gunsight: This was a major contributor to markedly improved shooting accuracy which removed the need for the pilot to estimate deflection angles, the new sight performing the necessary calculations for him and increasing the chance of scoring hits. The ability to judge deflection angles was a major factor in separating the top notch fighter pilots from the rest – the gyro gunsight meant that an average pilot could achieve a much greater level of success.

Introduced near the end of 1943 after four years' development at the Royal Aircraft Establishment (RAE) Farnborough, the gyro gunsight automatically calculated the angles after a few seconds tracking the target. Interestingly, many of the top pilots – who could accurately judge the angle without mechanical help – refused to use the new sight, but for the average squadron pilot it was a godsend and kill rates improved substantially.

Bubble canopy: Rearward view was always of vital importance to a fighter pilot's chances of surviving, but like most fighters designed at around the time, the Spitfire had a substantial blind area to the rear because of the original canopy design. The introduction of the bubble canopy and cut down rear fuselage top decking on some models from December 1944 allowed almost 360deg visibility and was universally liked by pilots.

Increased Fuel Capacity: As has been the case with many British fighters over the years, the Spitfire lacked fuel capacity and efforts to increase it were quickly made. In the case of the Spitfire, it was designed as a point interceptor without the need for a long endurance, but operational needs changed as the war progressed and Spitfire squadrons were more often flying into Europe from their British bases. Overseas operations also often demanded longer ranges than the Spitfire was capable of achieving.

External fuel tanks were developed, but increasing the aircraft's internal fuel capacity was important. Space for extra fuel was found in several areas. More fuel capacity in the wings had been developed for the photo-reconnaissance versions but this was more difficult in the armed versions because of the wing space occupied by guns. Some combat versions carried additional fuel in the wing leading edges, but most development in this area was based around rear fuselage tanks, possible with some repositioning of equipment. Centre of gravity issues came into play with resultant lateral instability as a result, but the bottom line was greater range – although this never approached the remarkable figures achieved by other aircraft, notably the P-51 Mustang.

Internal fuel capacities ranged from 85imp gal (386 litres) in the Mks.I, II, early V and XII to a maximum of 187imp gal (850 litres) in some Mk.XVIs with fuselage ferry tanks. Most late war models – both Merlin and Griffon powered – had fuel capacities in the 110-140imp gal (500-636 litres) range. Jettisonable external tanks were made in several standard sizes for the Spitfire: 30imp gal (136 l), 45imp gal (205 l), 90imp gal (409 l) and 170imp gal (773 l).

E Wing: As the war progressed, the hitting power of the 0.303in Browning machine gun diminished as enemy aircraft gained improved armour protection, leaving the two 20mm cannon also carried by most Spitfires from 1941 to do most of the work. A few had four 20mm cannon, but from 1944 the hard hitting Mk.II 0.50in heavy machine gun (as fitted to many US aircraft) became available for installation in Spitfires.

Housed in what was called the 'E' wing, the new armament comprised the usual pair of 20mm cannon plus two of the 50-calibres, the result being a notably greater weight of fire than could be achieved with the B or C wings' two cannon and four 0.303in machine guns.

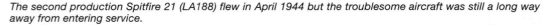

The second production Spitfire 21 (LA188) flew in April 1944 but the troublesome aircraft was still a long way away from entering service.

Two Seafire variants which had their first flights during the course of 1944: the Mk.XVII (top – a later FR.17 shown) in June; and the Mk.45 in September. TM379 shown here was the first Mk.45, built up from a Spitfire 21 airframe. Later in the year it was in the illustrated configuration with contra-props and an experimental fin and rudder shape.

The Spitfire PR.XIX entered service with 542 Squadron at Benson in May 1944. PM555 was a later production model, delivered a year later.

Bombing up.... an increasingly common sight at Spitfire units in the last year of the European war. This Mk.IX belongs to No 132 Squadron, operating from Normandy after D-Day. (via Neil Mackenzie)

After D-Day, one of the most important missions flown by Spitfires was the delivery of beer for the troops. In this case the liquid amber is being poured from the keg into a modified fuel tank. Subsequently, 'Depth Charge Modification XXX' was devised, allowing the kegs themselves to be carried under the Spitfires' wings!

ODDS and SODS

Some Spitfire oddities such as the floatplanes and proposed Plastic Spitfire have been covered elsewhere, but there were a few other ideas and concepts associated with the aeroplane which are worth a brief mention.

Many Spitfires were used for experimental purposes and to try out some bright ideas, among them: a Mk.II fitted with spatted fixed undercarriage; a Mk.V equipped for glider towing; two HF.VIIs modified for liquid oxygen injection tests; a PR.XI modified for mail carrying; and a 'civilianised' F.XIVE used by the Brazilian Air Attache.

One Spitfire IX was converted to a form of glider by Flight Refuelling Ltd in 1943. The idea was to extend the Spitfire's ferry range by towing it behind a Vickers Wellington bomber for a distance before starting the engine. Oiling up problems with the engine and propeller resulting from the long shut down time while the Spitfire was being towed were partially cured by fitting a feathering propeller but the excessive physical demands imposed on the pilot made the idea impractical.

The Malionowski Trailer was another range increasing idea for the Spitfire. It was basically a towed fuel tank in the form of a trailing boom with a horizontal tail surface and twin fins and rudders. The booms joined the Spitfire's wing just outboard of the cannon. The scheme wasn't tested on a Spitfire although a prototype smaller version was tried out on a Miles Magister trainer. The concept was developed by Polish fighter pilot and engineer Flt Lt Malionowski and taken up by Heston Aircraft Ltd which then proposed it to Supermarine in December 1942.

17 June: A Spitfire of No 602 Squadron is thought to be responsible for strafing a staff car carrying General Erwin Rommel near the front line in France. Rommel is injured and removed from the battle.

August: The first production Seafire XV (SR446) delivered from Westland.

15 August: Following several months of investigation into developing a Seafire based on the rewinged Spitfire 21, the Ministry of Aircraft Production directs that three Mk.21 airframes be set aside for as use prototypes for what would be called the Seafire Mk.45. This was subsequently reduced to two airframes, one of them becoming the Seafire 46 prototype. The jump in mark numbers resulted from the confusing proliferation of Spitfire/Seafire variants which had either appeared or were still planned, the decision having been made to restart the Seafire sequence from Mk.40. The Seafire XVII was originally to be called the Mk.41, and the stillborn Seafire XVIII with lengthened fuselage (the designation clashing with a Spitfire variant) the Mk.42.

18 August: Spitfire XIV RB144 fitted with a Griffon 85 engine and de Havilland six bladed contra-props.

30 August: Spitfire 21 LA192 (the sixth production aircraft) becomes the first of the mark to be handed over to the RAF, albeit for evaluation purposes. It would be more than seven months before the Mk.21 is finally cleared for operational service.

31 August: Completion of the prototype Seafire F.45 (TM379), built up from a Spitfire 21 airframe. As was the case with all Seafire 45s, this aircraft was fitted with a standard Spitfire 21 non folding wing.

September: Testing of Merlin 66 powered Spitfire LF.IX MJ636 with interconnected throttle and propeller controls. Late production aircraft featured this system, originally developed to counter supercharger stall problems which emerged if aircraft with two-speed/two-stage Merlins were flown at extremes of high boost and low revs at altitude. From now, the vast majority of Spitfire IXs built are of the LF.IX model with the low rated Merlin 66.

8 September: First flight of the prototype Seafire 46 (TM383).

October: First deliveries of the Spitfire XVI (first aircraft PV288), externally indistinguishable from late production Mk.IXs but with a Packard-Merlin 266 engine – equivalent to the British Merlin 66 with some modifications to the supercharger and cooling systems. Early operations included dive bombing the V-2 missile sites in Holland (from November 1944) with 250lb (113kg) or 500lb (227kg) bombs. The squadrons involved were Nos 229, 453 (RAAF), 602 and 603, and the diving technique was necessary due to the very small target area. The dive was started at 8,000ft (2,440m), achieved an angle of 70 degrees and the bombs were released at 3,000ft (915m).

October: The prototype Seafire 45 (TM379) fitted with a Rotol and then de Havilland six bladed contra-prop for testing.

October: Air Defence of Great Britain (ADGB) – established by renaming Fighter Command in November 1943 – reverts to its previous evocative title after a period of enormous unpopularity with RAF members and the

THE STILLBORN Mk.23

One Spitfire variant which did not achieve production was the Type 372 Mk.23 which would have been called the Valiant had it been manufactured. Externally, the Mk.23 as proposed was based on the Mk.21 but with a revised wing section with a raised leading edge and many of the characteristics of a laminar flow wing.

Two wing designs incorporating laminar flow principles were proposed, one an interim type (for the Mk.23/Valiant) and the other with a true laminar flow section as later applied to the Spiteful and Seafang fighters.

The interim wing was fitted to Spitfire VIII JG204 in early 1944, testing revealing performance improvements but less pleasant handling characteristics. The prototype Spitfire 21 (PP139) was also converted to something resembling the proposed Mk.23/Valiant but the arrival of the prototype very high performance Spiteful in June 1944 put paid to the Spitfire 23 project.

public. The original decision to revive the ADGB name (which had been used prior to 1936) was seen as a major mistake by the Air Ministry and one which had a detrimental effect on morale.

October: A Seafire IIC is tested in comparative trials with a captured Mitsubishi A6M5 Model 52 Zero ('Zeke') in the USA. Its conclusions: "Never dogfight with a Zeke 52 – it is too manoeuvrable. At low altitudes, where the Seafire is at its best, it should make use of its superior rate of climb and speed to obtain a height advantage before attacking. If jumped, the Seafire should evade by using superior rate of roll. The Zeke cannot follow high speed rolls and aileron turns."

5 October: The first shooting down of a Messerschmitt Me 262 jet fighter by Spitfires when five Mk.XIVs from No 411 (RCAF) Squadron successfully took one on over Nijmegan in the central Netherlands. On 25 December Flt Lt J J Boyle of the same squadron individually shot down two Me 262s in dogfights.

November: Formation of a special UK based Wing comprising four squadrons equipped with Spitfire LF.XVIEs – Nos 229, 453 (RAAF), 602 and 603 – to attack V-2 missile sites in Holland. The Spitfires were armed with two 250lb (113kg) or one 500lb (227kg) bomb and the attacks were largely ineffective. They continued into early 1945, and many sorties were abandoned due to poor weather.

November: The Spitfire FR.XIV fighter-reconnaissance aircraft enters service with Nos 2 and 430 Squadrons of the 2nd Tactical Air Force in France.

November: SX549, originally ordered as a prototype Spitfire 21 is built as the first Mk.22.

December: 'Modification 963' – the bubble canopy and cut down rear fuselage top decking – introduced to the Spitfire production line at Castle Bromwich for the Mk.XVI and late model Mk.IXs. Testing of the new canopy/fuselage arrangement had begun in mid 1943 but its early introduction would have disrupted mass production and it was delayed.

MERLIN and GRIFFONS

Crucial to the success of the Spitfire and Seafire, the Rolls-Royce Merlin and Griffon engines were developed to a remarkable degree during the course of the war, the Merlin's power increasing by some 70 per cent over the years (and more than doubling with the use of special fuels and very high boost pressures), all without changing the rev limit (3,000rpm) or the compression ratio (6.0:1).

What did change was the allowable boost pressures and the superchargers, starting with single-speed/single-stage units on the earlier versions and progressing through two-speed/single-stage and finally two-speed/two-stage.

These allowed the supercharger gear ratios to be 'tailored' to meet specific requirements, some engines producing maximum power at low altitudes, others at medium altitudes and still others at high altitudes. The larger capacity Griffon was similarly developed.

As they applied to the Spitfire, maximum boost pressures increased from +6$\frac{1}{4}$lb originally using 87 octane fuel and increasing to +12lb, +15lb, +16lb and +18lb with 100 octane fuel, each increase improving the power output. Some special Merlins used 150 octane fuel and up to +25lb boost, the engines – called 'Basta', see separate

The Rolls-Royce Merlin 61, as described by the detailed specification.

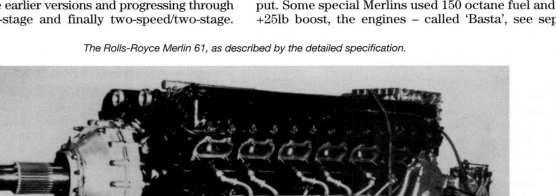

box – producing some 2,000hp (1,491kW). The same idea was applied to the Griffon.

All Merlins shared the same major characteristics: 60 degree liquid cooled V12 (cooling was 30 per cent glycol and 70 per cent water except the earliest models with 100 per cent glycol); bore and stroke 5.4in x 6.0in (137.16 x 152.40mm); capacity 27.0 litres (1,647cu in); compression ratio 6.0:1. Differing carburettors, cooling systems, supercharger configurations, propeller reduction ratios and so on were the main changes to the basic engine.

Generally speaking, Merlins up to and including the 50 series featured single-speed/single-stage superchargers, exceptions being the X and XX models with two-speed/single-stage blowers. The 60 and 70 series engines had two-speed/two-stage superchargers, the 60s featuring a variety of altitude ratings and the 70s generally for high altitude operations. Engines with an 'M' suffix in their designations (eg the 54M, 50M and 55M for the Spitfire

LF.V) had a cropped supercharger impeller which allowed maximum power to be developed at very low altitudes.

The Griffon shared the same 60 degree liquid cooled V12 configuration but was of about one-third greater displacement, its bore and stroke of 6.0in x 6.6in (152.40 x 167.64mm) giving a capacity of 36.7 litres (2,240cu in).

The earlier Griffons (Mks.III, IV, VI etc) had two-speed/single-stage superchargers, and the 60s, 70s and 80s had two-speed/two-stage blowers. The Griffon 80s were intended to drive counter-rotating propellers, while the later Griffon 100 engines (which powered no production Spitfire or Seafire) had two-speed/three-stage superchargers.

Merlin production reached 168,040 by Rolls-Royce Derby (32,377), Rolls-Royce Crewe (26,065), Rolls-Royce Glasgow (23,647), Ford Manchester (30,428) and Packard in the USA (55,523).

SPECIFICATION – MERLIN 61

Type: 60deg V12, liquid cooled.
Power: 1,290hp (962kW) for takeoff at 3,000rpm; 1,565hp (1,167kW) at 12,250ft (3,735m); 1,390hp (1,036kW) at 23,500ft (7,160m).
Max boost: +15lb.
Propeller reduction gear ratio: 0.42:1.
Compression ratio: 6.0:1.
Dry weight: 1,640lb (744kg).
Features: Medium/high altitude rating, electric starting, provision for cabin blower.
Application: Spitfire F.IX.
Cylinders: Bore 5.4in (137.16mm), stroke 6.0in (152.40mm), swept volume 1,647cu in (27.0 litres), cylinder blocks cast in RR50 aluminium alloy, high carbon steel cylinder liners, valve guides cast iron (inlet) and phosphor bronze (outlet).
Pistons: Machined from RR59 alloy, three compression and two scraper rings, fully floating hollow gudgeon pins of hardened nickel-chrome steel.
Connecting rods: H-section nickel steel forgings, plain and forked rod per pair, latter with nickel-steel bearing block, steel backed lead bronze alloy bearings.
Crankshaft: One piece six-throw, machined forging of chrome-molybdenum steel, nitrogen hardened, drive to reduction gear pinion through splined coupling shaft.
Crankcase: Halved, cast in aluminium alloy, upper part contains cylinders, crankshaft bearings and propeller reduction gear housing, lower portion sump case with oil pumps and filters, main

bearing split mild steel shells lined with lead-bronze alloys, bearing held in position by caps.
Wheelcase: Aluminium casting, supercharger unit attached to rear of wheelcase which houses drives to camshafts, magnetos, coolant and oil pumps, supercharger, hand and electric starters, and generator.
Valve Gear: Two exhaust and two inlet valves of KE95 steel per cylinder. Two concentric coil springs per valve, one steel rocker per valve, seven bearing camshaft per cylinder block.
Induction: Twin choke Rolls-Royce/SU updraught carburettor, automatic mixture control, twin fuel pumps.
Supercharger: Two-speed/two-stage, speed change operated by oil pressure, supercharger delivery pressure controlled by automatic servo mechanism coupled through differential linkage to throttle so that opening of throttle is controlled to suit boost pressure.
Ignition: Two 12-cylinder magnetos spigot mounted on each side of wheelcase, skew gear driven.
Lubrication: Dry sump system, one pressure and two scavenge pumps.
Cooling system: Coolant 70% water and 30% ethylene glycol, centrifugal pump, pressurised maximum permissible coolant temperature 135decC.
Intercooler: Coolant 70% water and 30% ethylene glycol, circulated by centrifugal pump from header tank through radiator to jacket between supercharger stages, on to intercooler matrix between supercharger and induction pipe and then back to header tank.

Before and after: the first production Spitfire F.22 (PK312) as it first appeared in early 1945 (top) with Mk.XIV style tail surfaces; and later in the year with the definitive Spiteful tail.

January 1945: After numerous delays caused by development problems mostly associated with handling, the Spitfire 21 enters squadron service with No 91 Squadron at Manston but it would be another three months before the squadron was declared operational on the new aircraft. No 91 was the only squadron to fly the Mk.21 operationally before the European war ended, undertaking 154 operational sorties in the month before hostilities ceased. It flew armed reconnaissance sorties over the Netherlands and against German miniature submarines operating off the coast of Holland and Belgium, claiming one sunk (see 26 April).

January: The first of 50 production Seafire 45s (LA428) completed at Castle Bromwich.

March: Testing of the first Spitfire XVI with bubble canopy, cut down rear fuselage decking and rear fuselage fuel tank (SM410). The aircraft was also fitted with metal elevators.

March: The Royal Navy's 21st Carrier Group comprising five aircraft carriers begins operations in the Bay of Bengal, largely in support of Allied landings and on shipping and air-ground support missions. The Group has ten fighter squadrons including three with Seafires – No 807 (aboard HMS *Hunter*), No 809 (HMS *Stalker*) and No 879 (HMS *Attacker*).

March: Manufacture of the Spitfire XVIII begins, this model replacing the Mk.XIV on Supermarine's production

The Spitfire F.21 finally entered service with No 91 Squadron at Manston in January 1945 but it would be another three months before the type was declared operational. In this shot, two F.21s flank the first production F.22.

lines. There was no prototype as such, the first aircraft (SM843) built to production standards.

21 March: Maiden flight of the first production Spitfire 22 (PK312).

April: The first production Seafire XVII (SW986) completed by Westland which built 212 between then and October 1946. Cunliffe-Owen contributed 20 to the production tally, starting from late December 1945 with SP323.

1 April: The Seafire claims its first air combat 'kills' in the Pacific when Lt R Reynolds of 894 Squadron aboard HMS *Indefatigable* shoots down three Zeros. Reynolds was the Fleet Air Arm's only Seafire 'ace'

having previously claimed two Bv 138 reconnaissance flying boats over the Atlantic. *Indefatigable* was one of four British aircraft carriers attached to the American Fifth Fleet's Task Force 57, whose carriers, battleships, cruisers and destroyers supported the Okinawa landings in late March 1945 and other operations.

12 April: The Spitfires of the 2nd TAF's No 125 Wing become the first to cross the Rhine, establishing a base at Twente.

20 April: The first Spitfire 22 delivery to the RAF (PK313).

26 April: The only Spitfire 21s to see operational service during the war – those of No 91 Squadron based at

Seafire XVII deliveries began in April 1945 followed by operational service five months later.

SCRATCH ONE JET

On 2 March 1945, Flt Lt Danny Reid DFC (an Australian flying with No 41 Squadron RAF) had an encounter with an Arado Ar 234 twin engined jet reconnaissance bomber near Enschede. The Ar 234 had just entered limited service with the *Luftwaffe* and the Allies knew very little about it. As the squadron's artist, Reid was able to draw a very accurate sketch of the aircraft when he returned from the mission despite having to work entirely from memory and having never seen an Ar 234 before.

Reid was flying Spitfire XIV RM796 and his combat report is reproduced below:

"I took off from B80 [a Belgian landing field] at 0702 as Blue 1 with Flt Sgt Kelly as Blue 2 to patrol Nijmegen at approx 0735..... reported a jet aircraft NE of Weert. I flew towards Weert in cover of thin layers of cloud, then returned towards Nijmegen still between cloud layers. On emerging near Nijmegen saw a single aircraft about one mile in front and about 1,000ft above. I immediately gave chase instructing my No 2 to slow down and drop his tank, being unable to do so myself due to the high speed.

"The E/A [enemy aircraft] turned slightly to starboard and continued towards the north-east, weaving slightly from time to time. I kept out of the enemy pilot's view by keeping under his tailplane and slowly overhauled him at an IAS [indicated airspeed] of 340mph at 8/9000ft. Whilst astern of the E/A I was only able to say it was a jet propelled and not an Me 262 or Meteor.

"I closed to 100 yards or less, firing with 0.5 MG and cannon whilst still overtaking. I saw strikes on the port wing, port jet engine and fuselage. E/A immediately emitted dense clouds of brownish black smoke, possibly jet exhaust. I continued firing and saw flashes in the smoke, breaking away at extremely close range and being hit in the port radiator by debris. I next saw the E/A going down in a wide spiral to starboard with smoke or vapour pouring from holes all along the port wing, and dark smoke from the fuselage.

"I could then see the long nose of the a/c and straight tapered wings and identified E/A as an Arado 234. A large piece of E/A then flew off, and one person bailed out, parachute opening. E/A steepened its dive and crashed somewhere near Enschede, being completely destroyed by explosion. Some fire on the ground was seem at this time.

"I returned to patrol being rejoined by my No 2 who witnessed the destruction of this enemy a/c. I claim one Arado 234 destroyed.

Late production Spitfire LF.XVIEs at Castle Bromwich in the first half of 1945. The last was delivered in July 1945, marking the end of the Merlin powered Spitfire line.

SUMMARY OF CONTRACTS

The following table summarises the Spitfire and Seafire production contracts placed over the years, showing the date, quantity, type and manufacturer. Many of these orders were substantially amended and/or cancelled, especially in the later war years, so they do not necessarily reflect what was actually built. The idea is to show the rapid pace at which orders were placed and the large numbers involved. Refer to the serial numbers table for information on how these contracts were amended to meet rapidly changing needs.

Date	Manufacturer	Qty	Model/s
6/36	Supermarine	310	Spitfire I
9/38	Supermarine	200	Spitfire I
4/39	Castle Bromwich	1000	Spitfire II
4/39	Supermarine	200	Spitfire I
8/39	Supermarine	450	Spitfire I
3/40	Supermarine	450	Spitfire I
6/40	Supermarine	500	Spitfire I
7/40	Supermarine	500	Spitfire I
6/40	Castle Bromwich	500	Spitfire I
6/40	Supermarine	2	Spitfire PR Type D
8/40	Westland	300	Spitfire I
10/40	Supermarine	1100	Spitfire IA/B
10/40	Castle Bromwich	1000	Spitfire III
10/40	Supermarine	120	Spitfire III
8/41	Supermarine	500	Spitfire VC
8/41	Castle Bromwich	904	Spitfire V
8/41	Castle Bromwich	750	Spitfire IV
9/41	Westland	200	Spitfire IV
10/41	Westland	185	Spitfire V
1/42	Supermarine	800	Spitfire VIII
1/42	Castle Bromwich	989	Spitfire VB
2/42	Castle Bromwich	680	Spitfire VC
3/42	Supermarine	202	Seafire IIC
3/42	Westland	213	Seafire IIC
3/42	Castle Bromwich	300	Spitfire VC
4/42	Supermarine	70	PR Spitfire (Merlin 61)
5/42	Supermarine	426	Spitfire VC
5/42	Castle Bromwich	2190	Spitfire VC
6/42	Supermarine	592	Spitfire VIII
7/42	Supermarine	260	Seafire IIC
7/42	Supermarine	700	Spitfire VIII
11/42	Westland	200	Seafire IIC
12/42	Castle Bromwich	600	Spitfire IX
12/42	Supermarine	225	Spitfire VIII
12/42	Supermarine	144	Spitfire VIII
1/43	Supermarine	60	Seafire IIC
1/43	Westland	200	Seafire IIC
1/43	Cunliffe-Owen	250	Seafire III
4/43	Supermarine	100	Spitfire VIII
6/43	Castle Bromwich	800	Spitfire IX
6/43	Castle Bromwich	1500	Spitfire 21
7/43	Supermarine	200	Spitfire VIII
7/43	Westland	250	Seafire III
7/43	Cunliffe-Owen	150	Seafire III
8/43	Supermarine	50	Spitfire XIV
8/43	Supermarine	373	Spitfire IX
10/43	Castle Bromwich	100	Spitfire IX
10/43	Supermarine	406	Spitfire XIV
11/43	Castle Bromwich	73	Spitfire IX
1/44	Castle Bromwich	700	Spitfire 21
1/44	Westland	300	Seafire III
2/44	Castle Bromwich	800	Spitfire 21
2/44	Supermarine	6	Spitfire VII
2/44	Westland	140	Seafire XV
2/44	Supermarine	150	Spitfire PR.XI
2/44	Cunliffe-Owen	200	Seafire XV
3/44	Westland	503	Seafire XV
4/44	Castle Bromwich	1884	Spitfire IX
7/44	Castle Bromwich	276	Spitfire IX
8/44	Supermarine	300	Spitfire XIV
2/45	Supermarine	157	Spitfire XIV
3/45	Castle Bromwich	188	Spitfire XVI
3/45	Castle Bromwich	50	Spitfire 22
3/45	Supermarine	100	Spitfire PR.XIX
4/45	Supermarine	600	Seafire 45/46/47
11/45	Castle Bromwich	150	Spitfire 22
11/45	Supermarine	135	Seafire 46/47
4/46	Supermarine	64	Seafire 47
10/46	Supermarine	100	Seafire 47

Ludham in Norfolk – achieve success of a kind when two of them strafe and claim sunk a German midget submarine caught on the surface. The pilots were Flt Lts J Draper and W Marshall.

30 April: The last major air battle in the war against Germany when Spitfires and other Allied fighters take on Fw 190s and Bf 109s, destroying 37 of them. The action took place during the Allied crossing of the Elbe. Hitler killed himself in Berlin on the same day.

May: The Seafire XV enters RN FAA service with No 802 Squadron.

8 May: VE Day.

28 May: The first Spitfire XVIII delivered to the RAF (FR.XVIII SM844). The aircraft was passed on to 28 Squadron in Hong Kong in July 1945. With hostilities winding down, many of the 300 Spitfire XVIIIs built up to January 1946 were delivered to storage, the remainder going to squadrons based in India and the Far East.

31 May: A 'stock take' of the RAF's aircraft inventory

Seafire XVII SX235 lands on HMS Illustrious. *The aircraft was just too late to see World War II service, becoming operational with the Fleet Air Arm's No 883 Squadron in September 1945.*

reveals that 5,864 Spitfires are held on strength. Of these, only 510 are found to be unsuitable for immediate use.

June: The first Spitfire XIVs for Far East service delivered to No 11 Squadron based at Chettinad in India. The aircraft were to be used in support of the planned Allied invasion of Malaya, but Japan's surrender meant this operation was not necessary. The squadron went to Malaya three months later and in May 1946 took its Spitfire XIVs and XVIIIs to Japan as part of the Commonwealth occupation force.

July: The end of Merlin-Spitfire production with the delivery of Mk.XVI TE480.

27 July: The Royal Navy 21st Carrier Air Group's Seafires record their first air-to-air 'kills' when the fleet is attacked by seven Japanese aircraft on a *Kamikaze* mission in the Bay of Bengal. The Seafires claim three, two were shot down by anti aircraft fire, one hit and sank the minesweeper HMS *Vestal* and the other slightly damaged the carrier HMS *Ameer*.

6 August: Atomic bomb dropped on Hiroshima.

9 August: Atomic bomb dropped on Nagasaki.

14 August: Japan agrees to unconditional surrender.

September: The Seafire XVII enters service with the Fleet Air Arm's No 883 Squadron.

13 September: Less than two weeks after the re-occupation of Hong Kong by Britain, the aircraft carrier HMS *Smiter* arrives at the Colony with the Spitfire XIVs of No 132 Squadron RAF aboard. The Spitfires flew off the small carrier with some difficulty – there was no catapult launching facility – but did so successfully and established themselves at Kai Tak. The squadron remained in Hong Kong for only seven months before disbanding, flying mainly anti piracy patrols in the meantime.

November: First production Seafire 46 (LA541) completed. Production of only 24 examples was at a leisurely pace with the last flown in March 1947.

Spiteful F.XIV RB520 fitted with an arrester hook to serve as the prototype for the naval Seafang.

SPITEFUL and SEAFANG

The quest for increased speeds in fighters – in both the dive and level flight – led to the development of more advanced aerofoils in what was a period of transition between the piston engine/propeller combination and the jets which would start appearing from the mid 1940s.

British Air Ministry Specification F.1/43 led to the Supermarine Type 371 Spiteful, which built on late model Spitfire technology but featured a laminar flow wing which achieved maximum thickness at 42 per cent chord compared to 25 per cent on the Spitfire. The new wing had straight leading and trailing edges, completely losing the original Spitfire elliptical shape.

The Rolls-Royce Griffon provided the power in several versions for the planned production models: 2,375hp (1,771kW) Griffon 69 and five bladed propeller (F.XIV); 2,350hp (1,752kW) Griffon 89/90 and six bladed contra-props (F.XV); or 2,420hp (1,805kW) Griffon 101 and five bladed propeller (F.XVI), the latter with three-stage supercharger.

Three prototypes were ordered in February 1943 with 60 production examples ordered four months later but these were cancelled. Another order for 21 aircraft was placed, of which 17 were built. The first prototype (NN660) was basically a Spitfire XIV fuselage mated with the Spiteful's new wing and undercarriage. It first flew on 30 June 1944 but crashed during September, killing test pilot Frank Furlong.

The second aircraft (NN664) flew in January 1945 as the first true Spiteful but serious handling problems relating to aileron snatch, wing drop before the stall and a

powerful flick during a high 'g' stall remained difficult to solve. The fitting of the definitive larger horizontal and vertical tail surfaces (also applied to the Spitfire 22/24) helped directional stability but the problems were never fully solved and the Spiteful was declared obsolete for service purposes in October 1947.

Seventeen production Spitefuls were built, most to F.XIV standards, a few as F.XVs and one (RB518) as an F.XVI with Griffon 101. The aircraft was certainly fast, the F.XIV achieving a maximum speed of 483mph (777km/h) at 21,000ft (6,400m), while the F.XVI recorded the fastest level speed achieved by a British piston engined aircraft – 494mph (795km/h) at 28,500ft (8,685m).

A naval version of the Spiteful was formally approved at the end of 1944 under the name Seafang and planned in two versions: the F.31 with non folding wings, Griffon 69 and five bladed propeller and F.32 with folding wings, increased internal fuel capacity, Griffon 89/90 and contra-props. Like the Spiteful, the Seafang did not enter quantity production, the order for 150 being cut with the result that only 10 F.31s and eight F.32s were built, of which several delivered were dismantled and never flown. The Royal Navy preferred to stick with the last Seafire models.

Spiteful F.XIV RB520 was partially 'navalised' to serve as the prototype Seafang and the first production standard aircraft appeared in early 1946. Production and development was protracted as the Royal Navy's requirements changed, the service finally deciding to order its first jet fighter – the Supermarine Attacker – instead.

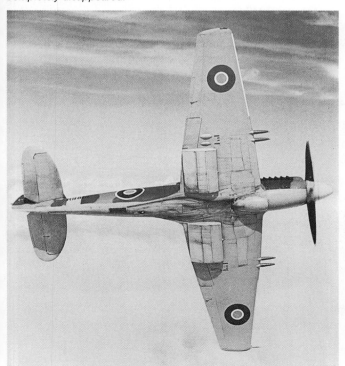

An underside view of a Spiteful clearly shows the planform of the new laminar flow wing. The classic Spitfire elliptical wing shape has completely disappeared.

(above) The first prototype Seafang F.31 (VG471). (below) VB895, the prototype Seafang F.32 with contra-props and folding outer wings.

The last of the line – the Seafire FR.47, first flown on 25 April 1946. These aircraft are from No 800 Squadron FAA and were photographed aboard HMS Triumph as it sailed from Malta in 1947.

February 1946: The first of 48 Seafire IIIs transferred to the French Navy.

27 February 1946: First flight of prototype Spitfire 24, converted from Mk.22 PK313.

25 April 1946: Maiden flight of the first Seafire 47 (PS944), the ultimate development of the Spitfire/Seafire line. All but the first four of 90 Mk.47s had hydraulically powered wing folding and unfolding. The first 16 aircraft were powered by the 2,145hp (1,600kW) Griffon 87 but the remainder had the 2,350hp (1,752kW) Griffon 88 with a Rolls-Royce developed fuel injection system instead of the previous Bendix-Stromberg fuel metering carburettor. All had six bladed Rotol contra-props.

May 1946: First flight of the first 'official' two seat Spitfire, converted from the LF.VIII MT818. The aircraft had originally first flown in May 1944 and after being used for trials, was sold to Vickers-Armstrong and converted to a Type 502 Spitfire Trainer with the normal cockpit moved 13.5in (34.3cm) forward and a second, raised cockpit inserted behind. Moving the front cockpit forced a decrease in the capacity of the forward fuselage fuel tank. The aircraft first flew with the Class B registration N32 but was then re-registered G-AIDN. Twenty other Spitfire Trainer conversions were subsequently performed, all from Mk.IXs (Type 509). Even though the idea of a two seat Spitfire had been mooted as early as 1941, only two

unofficial and much more basic field conversions are believed to have been performed before G-AIDN's appearance – a Mk.VC (ES127) flown by No 261 Squadron based in Sicily in 1944, and a few Mk.IXs by the Russians during the war.

31 August 1946: The Irish Air Corps orders 12 ex FAA Seafire IIIs for land based operations, these aircraft 'denavalised' and converted to effectively Spitfire VC standards. They were delivered between January and September 1947.

November 1946: Spitfire FR.XIVE TZ138 sent to Canada for winterisation trials and is damaged when the wheels sink into the snow, causing the aircraft to nose

The first Spitfire F.24 flew on 27 February 1946. This is VN324, built five months later.

PK664, one of the small number of Spitfire F.22s fitted with contra-props.

over onto its propeller. A set of Tiger Moth skis was then fitted, these dropping away when the aircraft left the ground. A very high takeoff speed was required with the skis fitted (120-150mph/193-240km/h) and the report on the trial not surprisingly recommended that single engined fighters should not operate from skis except in an emergency.

5 May 1947: A civil registered Spitfire PR.XI (LV-NMZ) purchased for aerial survey and mapping work in Argentina completes a ferry flight from Hurn (England) to Buenos Aries (Argentina) via Gibraltar, Dakar, Natal, Rio de Janeiro and Montevideo flown by Capt James Storey. The 1,860 miles (2,993km) between Dakar (Senegal) and Natal (Brazil) was the longest flight by any Spitfire, taking $8\frac{1}{2}$ hours to complete and accompanied by a British South American Airways Avro York. Fuselage and wing tanks plus a 170imp gal (773 l) underfuselage overload tank gave the aircraft a total fuel capacity of 428imp gal (1,945 l), sufficient for a theoretical still air range of over 2,200 miles (3,540km).

13-15 May 1947: Seafire 47 deck handling trials aboard HMS *Illustrious* using PS945, PS948 and PS 949.

18 August 1947: The first of three Spitfire PR.XIs delivered to the Royal Danish Air Force (ex RAF PL794, PL833 and PM134).

January 1948: No 80 Squadron based at Gutersloh in Germany replaces its Hawker Tempest Vs with Spitfire

Spitfire F.24 production at South Marston in 1947.

SPITFIRE VERSUS SPITFIRE

There are two recorded examples of intentional Spitfire versus Spitfire combat, both of them during the Arab-Israeli war of 1948-49. The first occurred on 22 May 1948 when an Egyptian Spitfire IX bombed Ramat David in Palestine where two RAF squadrons of Spitfire FR.18s (Nos 32 and 208) were covering the British withdrawal from the area, a week after the British Palestinian Mandate ended and the establishment of the State of Israel.

Two British Spitfires were set alight by the attack and the raider escaped despite four RAF aircraft being scrambled. Another attack was made two hours later by three Egyptian Spitfires with more damage resulting. This time, the RAF FR.18s were ready and successfully intercepted. All three Egyptian Spitfire IXs were destroyed, one of them by ground fire. The Egyptian Government put the attacks down to "a regrettable navigation error".

The Spitfire's air-to-air combat career approached its conclusion with a remarkable incident on 7 January 1949. On that date, four Spitfire FR.18s of the RAF's No 208 Squadron were fired on by Israeli anti aircraft guns while conducting a low level reconnaissance near the Israeli frontier.

One was hit and its pilot bailed out while the other three were 'bounced' by two Spitfire IXs of Israel's No 101 Squadron – and all three RAF Spitfires were shot down, Israel claiming the RAF aircraft had infringed its airspace. To add insult to injury, one of the four RAF Hawker Tempests sent out to search for the missing Spitfires was also shot down by an Israeli Mk.IX.

24s and becomes the only RAF operational squadron to operate the final Spitfire mark.

20 February 1948: The last Spitfire – F.24 VN496 – leaves Supermarine's South Marston production line and is handed over to the Royal Air Force.

23 March 1948: The first three Type 509 Spitfire IX Trainers (T.9) delivered to the Royal Netherlands Air Force. The T.9 conversion was similar to the original T.VIII with the front cockpit moved slightly forward and a second, raised cockpit installed behind. Other customers were India (10), Egypt (1) and the Irish Air Corps (6), for a total of 20.

15 May 1948: The State of Israel comes into being and two Royal Egyptian Air Force Spitfire LF.IXs attack Dov airfield in Tel Aviv. One is forced to make a belly landing on a beach after being hit by anti aircraft fire and despite being badly damaged is repaired to enter service as the Israeli Air Force's first Spitfire.

June 1948: The Israeli Government concludes negotiations with Czechoslovakia to purchase 50 Spitfire IXs being phased out of service by the Czech Air Force. Deliveries began in September 1948, with 56 Spitfires eventually coming from that source.

6 July 1948: Two Spitfire FR.18s from No 60 Squadron based at Kuala Lumpur conduct the type's first combat mission of Operation *Firedog*, the lengthy campaign

A Seafire XV photographed in July 1947 demonstrates its wing folding system.

against communist terrorists during the Malayan Emergency. The action was against a terrorist camp in Perak, the Spitfires using cannon and rocket projectiles to destroy the target. The Spitfire was banned from carrying rockets and bombs in Malaya shortly afterwards when faulty wiring in one of them caused the inadvertent release of a rocket on the ground, killing a civilian. Spitfire strikes continued, but with cannon only.

28 January 1949: The 22,759th and last of the Spitfire/Seafire line delivered, a Seafire F.47 (VR972) for the Royal Navy Fleet Air Arm.

11 April 1949: The Spitfire VIII Trainer prototype conversion (G-AIDN, ex MT818) is demonstrated to the RAF to fill a possible requirement for about 20 aircraft to fill the gap between the Harvard and Spitfire 22 of the Auxiliary Squadrons. No order was placed and G-AIDN remained the only T.VIII conversion, although 20 Spitfires were converted from Mk.IXs.

May 1949: No 28 Squadron equipped with Spitfire FR.18s deploys to Kai Tak in Hong Kong from Malaya. It is joined in August by No 80 Squadron (equipped with Spitfire 24s) to form a new Wing .

Spitfire FR.14E TZ138 on Tiger Moth skis during winterisation trials in late 1946.

The long range civil registered Spitfire PR.XI purchased for aerial survey work in Argentina in 1947.

October 1949: No 800 Squadron Fleet Air Arm takes the ultimate Seafire variant, the Mk.47, into action for the first time. Embarked on HMS *Triumph* at Singapore but operating from dry land (RNAS Sembawang), the squadron's Seafires mounted rocket attacks against communist terrorist bases. No 800 was the only FAA squadron to take the Seafire 47 into action, including in the Korean War.

1 May 1950: A contract signed for the supply of 20 ex RAF Spitfire 22s to the Egyptian Government.

4 July 1950: The Seafire 47s of 800 Squadron FAA aboard HMS *Triumph* make their first rocket attack of the Korean War against the airfield at Haeju. The carrier and its Seafires were withdrawn to Hong Kong late in the following September with only one aircraft remaining fully serviceable. The squadron had flown 360 combat air patrol, ground attack and anti shipping (mostly against suspect junks) operational sorties in the meantime. One Seafire had been shot down by a 'friendly' USAF B-29 and other losses were attributable to accidents, but a shortage of spare parts was the main reason for 800's lack of serviceable aircraft.

November 1950: HMS *Triumph* and 800 Squadron's Seafire 47s return to Britain after Korean War service and the squadron is disbanded, thus ending Royal Navy front line service of the aircraft. Training and reserves units continued operating the aircraft for another four years.

1 January 1951: No 60 Squadron, based at Tengah, flies the RAF's last Spitfire combat sortie when its FR.18s

SPITFIRE VERSUS LIGHTNING

A highly unusual dissimilar air combat trial took place in 1963 between Spitfire PR.XIX PS853 and English Electric Lightning F.3 XP695. The Spitfire had been withdrawn from service in late 1957 and used as a 'gate guardian' at RAF Binbrook but maintained in an airworthy condition.

It was resurrected for the trials, conducted during the period of Confrontation with Indonesia when the possibility of RAF supersonic jets having to take on Indonesian P-51 Mustangs was a real one. The Spitfire was therefore selected to simulate a Mustang in tests against the Lightning and as such performed the last operational act by any of its type.

The trials revealed that the Spitfire had little chance against the Lightning unless the jet's pilot was silly enough to get into a turning and ever slowing dogfight. Even then, the Lightning still had the option of lighting the 'burners' and accelerating away, although it probably would have taken some hits from the more agile Spitfire in the meantime. Then there was the possibility of the Lightning running out of fuel....!

The Lightning's best option was to approach the Spitfire from behind and below in a high speed climb, providing a chance to get into a missile firing position without being seen.

PS853 Spitfire went on to fly with the Battle of Britain Memorial Flight.

Spitfire F.24 VN318, one of eight transferred to the Hong Kong Auxiliary Air Force in early 1952.

Dissimilar air combat training taken to the extreme. Spitfire PR.XIX PS853 and an English Electric Lightning ready for some mock combat in 1963 to devise tactics in case the Lightning had to take on Indonesian P-51 Mustangs during the period of Confrontation with that country.

conduct a strike against terrorists in the Kota Tinggi area of Johore during the Malayan Emergency. The squadron re-equipped with de Havilland Vampire FB.5s in the same month. During 18 months of front line service in Malaya, the squadron's Spitfires had flown some 1,800 operational sorties.

30 July 1951: The last of six Spitfire IX Trainers delivered to the Irish Air Corps, also the last of the 20 T.9s converted.

December 1951: The RAF's last front line fighter Squadron operating Spitfires (No 80 at Kai Tak in Hong Kong) begins re-equipping with de Havilland Hornets. The last of its Mk.24s is relinquished the following month. Eight of the aircraft were transferred to the Hong Kong Auxiliary Air Force: PK687, PK719, PK720, VN308, VN313, VN318, VN482 and VN485. No 81 Squadron at Seletar (Singapore) kept its last Spitfires until April 1954 (see below) but they were photo-reconnaissance versions.

1 April 1954: The last operational RAF Spitfire sortie when PR.19 PS888 of 81 Squadron based at Seletar (Singapore) conducted a photographic reconnaissance mission over some jungle in Johore, looking for communist guerrillas.

23 November 1954: With the disbandment of the Fleet Air Arm's No 764 (Training) Squadron at RNAS Yeovilton, Royal Navy Seafire service ends.

21 April 1955: Flt Lt Adrian Rowe-Evans of the Hong Kong Auxiliary Air Force performs the last sortie by a Spitfire 24 (VN485) during the Queen's Birthday Flypast.

12 June 1957: The final Spitfire flight in regular RAF service, PR.19 PS853 of the Temperature and Humidity Flight at Woodvale (Lancashire) conducting a meteorological sortie.

HIGHEST and FASTEST

The greatest speed ever attained by a Spitfire (in a dive) and the highest altitude both occurred during a single flight by Flt Lt Ted Powles of 81 Squadron based at Kai Tak (Hong Kong) on 5 February 1952. The aircraft was Spitfire PR.19 PS852.

On that day, Flt Lt Powles undertook a meteorological flight, his aim to record the outside air temperature at various set altitudes. After reaching an extraordinary 50,000ft (15,240m), Flt Ltd Powles decided to try for a bit more, finally reaching 51,550ft (15,710m) at which point there was little control left.

The cabin pressurisation warning light coming on necessitated a rapid descent to a safe altitude so as to prevent the pilot suffering 'the bends', but changing the aircraft's attitude produced a substantial loss of control which resulted in it entering a vertical dive. As speed increased, the effects of compressibility made normal control inputs ineffective.

Unable to see due to the canopy misting up, Powles finally regained control at about 2,000ft (610m), the descent from over 51,000ft having taking only 56 seconds! Some post flight calculations revealed the Spitfire had reached a true airspeed of 690mph (1,110km/h) during its descent, or Mach 0.94.

Spitfire VIIIs A58-315, 395, 405 and 409, four of the 410 of this model supplied to the Royal Australian Air Force between 1943 and 1945. The RAAF operated a total of 656 Spitfires 'at home' plus those flown by the four squadrons operating in the UK, Europe and Middle East under RAF control.

FOREIGN OPERATORS

FOREIGN OPERATORS

The RAAF received the first of 245 Spitfire VCs (and one VB) in August 1942, the aircraft originally destined to equip squadrons based in Australia's north to repel Japanese attacks. These aircraft still carry their original British serials, despite being allocated RAAF serials in the 'A58' series. For some inexplicable reason, the cannon stubs on the aircraft in this photograph (and others in the series) have been crossed out by someone. (via Neil Mackenzie).

Australia

The Royal Australian Air Force contributed four Spitfire squadrons (Nos 451, 452, 453 and 457) to the war effort in Europe and the Middle East between 1941 and 1945, operating under RAF control and flying Mks.I, II, V, IX and XIV. Two of the squadrons (452 and 457) returned to Australia to fight the Japanese in 1942.

The real threat of an attempted Japanese invasion of Australia in early 1942 prompted an urgent need for fighters for the RAAF, the USA responding quickly with P-40 Kittyhawks. Spitfires were also requested but these took longer to arrive largely due to Britain's own pressing needs.

In May 1942 it was agreed that three Spitfire squadrons would be allocated to assist in the defence of Australia. Two of these were Nos 452 and 457 which had been operating in Europe and the third was the RAF's No 54 Squadron. The three units were formed into No 1 ('Churchill') Wing operating in northern Australia.

Although the squadrons were established in Australia in August and September 1942, it wasn't until January 1943 that they were based in Darwin and operational. The tropicalised Spitfire V was selected initially, the first of 245 F.VCs (serial numbers A58-1 to -162, A58-164 to -185 and A58-200 to -259) plus a single F.VB (A58-163) arriving in August 1942. F.VC A58-259 was the last Spitfire V built at Castle Bromwich. Deliveries of 410 Spitfire F, LF and HF.VIIIs began in October 1943 (serials A58-300 to -550 and A58-600 to -758), although about 100 went directly to storage in 1945 as the Pacific War wound down.

There were some problems with the Spitfire's early Australian service, mainly due to its lack of range and tactical shortfalls. A May 1943 incident brought the issue to a head when 18 Japanese fighters and 24 bombers approached Darwin. No 1 Wing scrambled to intercept but did not reach the enemy until after Darwin had been bombed.

The battle which followed was fought at some distance from the Spitfires' base and as a result their meagre fuel reserves were stretched with five aircraft running out of fuel on the way home. From then, fuel gobbling dogfights were avoided, a 'hit and run' tactic adopted instead.

Other RAAF Spitfire squadrons 'at home' were Nos 79 and 85 and all eventually re-equipped with Mk.VIIIs. Nos 79, 452 and 457 were used in the 'island hopping' operations as Japan was gradually pushed back – although most activities were of a minor and in many ways wasteful 'mopping up' nature – while two RAF squadrons (Nos 548 and 549) remained in Darwin.

The end of the Pacific War saw the Spitfire's RAAF career come to an abrupt end and by early 1946 no fewer than 339 were in long term storage awaiting scrapping. A few were transferred to the Royal Australian Navy in 1948 as non flying ground instruction airframes and after three or four years were taken to fire dumps and burnt.

A Spitfire VC of No 54 Squadron RAF is pushed back into its camouflaged bay at Darwin after a sortie. This British unit operated under RAAF control during the defence of Australia's north. (via Neil Mackenzie)

The RAAF's No 452 Squadron returned home from Britain in mid 1942 to help defend Australia. Originally equipped with Spitfire VCs, it later changed to Mk.VIIIs, one of which is seen here being started at Sattler in the Northern Territory during August 1944. (via Neil Mackenzie).

Belgium

Two Belgian manned Spitfire squadrons (Nos 349 and 350) had operated as part of the RAF's 2nd Tactical Air Force in the latter part of the war and in October 1946 these were transferred to Belgian control as part of the *Force Aérienne Belge/ Belgische Luchtmacht* (Belgian Air Force). Ex RAF Merlin engined Spitfires were supplied pending delivery of FR.XIVs, these comprising 51

Mk.IXs and 25 Mk.XVIs and forming the basis of the service's post war rebirth. Fifteen Mk.IXs were also obtained from the Netherlands in 1950.

Deliveries of the Spitfire XIVs began in 1948, a total of 132 being received between then and 1951. After being retired from front line squadron service, the Spitfires were used by the Belgian Air Force's fighter school with the last retired in 1956.

Burma

With the independence of Burma in 1948 came the establishment of the Union of Burma Air Force. The service was built up with British assistance, 20 'denavalised' (by Airwork Ltd) Seafire XVs being delivered in 1951-52 followed by 30 Spitfire IXs purchased from Israel in 1955. Serial numbers in the UB400 range were carried and the Spitfires were replaced by Hawker Sea Furies in 1958-59.

One of 20 'denavalised' Seafire XVs supplied to the Union of Burma Air Force in 1951-52.

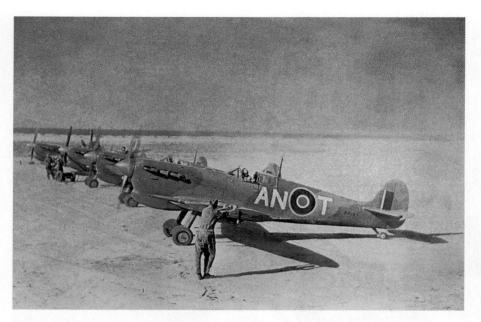

The Royal Canadian Air Force contributed 14 Spitfire squadrons to the RAF's forces. No 417 operated in the Middle East, its Spitfire VCs photographed here in Italy in early 1943.

Spitfire XIV RN119 of No 402 (RCAF) Squadron while operating with the 2nd Tactical Air Force in the European Theatre.

The Royal Canadian Navy acquired a number of Seafire XVs post war including PR479.

Canada

The Royal Canadian Air Force contributed 14 Spitfire squadrons to the European war effort, operating under RAF control. Nos 403 and 411 were the first, established in the UK in May and June 1941 equipped with Spitfire Is.

Subsequent squadrons were numbers 400, 401, 402, 412, 414, 416, 417, 421, 430, 441, 442 and 443, all operating in the European theatre except for No 417, which went to the Middle East, Western Desert and Italy equipped with Spitfire VIIIs. The others operated a variety of Spitfire marks as the war progressed including the Mks.II, V, IX, XIV and XVI. No 400 Squadron was a specialist photo-reconnaissance unit equipped with Spitfire PR.XIs.

Canadian units were prominent during the operations associated with the D-Day landings, with six home based fighter squadrons moving to Britain in preparation for Operation *Overlord*. Three of these were equipped with Spitfire Vs and IXs and renumbered as 441, 442 and 443 Squadrons for service as part of the 2nd Tactical Air Force.

Post war, the Royal Canadian Navy received Seafire 15s.

Czechoslovakia

The three Czech Spitfire squadrons (Nos 310, 312 and 313) which had operated as part of the RAF from 1941 returned to their homeland in the second half of 1945 to form the nucleus of the re-established Czech Air Force (*Ceskoslovenske Vojenske Letectvo*).

Seventy-three Spitfire LF.IXs were ferried to Prague to equip the units within three air regiments but when the communists seized power in 1948 Soviet types were introduced to service. As a result, about 60 of the Spitfires were sold to the newly formed state of Israel.

Denmark

The Spitfire was an important element of Denmark's post war re-equipment programme, the *Flyvevabnet* receiving 38 Mk.IXs and three PR.XIs in 1947 together with some non flying instructional airframes. They equipped three *Eskadrille* and with the establishment of the independent Royal Danish Air Force in 1950 they were used as both fighters (operating alongside

Gloster Meteors) and advanced trainers.

The serial numbers 401-438 were allocated to the Spitfire IXs and 451-453 to the PR.XIs. The last Danish Spitfires were retired in May 1955, replaced by Meteors and Republic F-84 Thunderjets.

Egypt

The Royal Egyptian Air Force received 15 Spitfire Vs before the war had ended followed by others from RAF stocks from October 1946. The final tally of Merlin powered Spitfires was 20 Mk.Vs, 41 Mk.IXs and one T.9.

Some of the Mk.IXs were involved in the first deliberate Spitfire versus Spitfire combat on 22 May 1948 over Palestine when the RAF base at Ramat David was attacked (see *Spitfire Versus Spitfire* breakout box) during the Arab-Israeli War.

Egypt also received 20 refurbished ex RAF Spitfire 22s in 1950. These had been largely replaced by de Havilland Vampire jet fighters by early 1955 and relegated to training duties.

France

The *Armée de l'Air* received a single Spitfire I for evaluation in July 1939 (the only pre war export) but in 1941 the first of three Free French Spitfire squadrons (Nos 340, 341 and 345) was formed in Britain as part of the RAF. Equipped progressively with Spitfire IIs, Vs, IXs and XVIs, they were transferred to the *Armée de l'Air* in November 1945.

Four other Free French squadrons were formed within the RAF (Nos 326, 327, 328 and 329) in 1943-44

(above) The Royal Egyptian Air Force received a total of 82 Spitfires including 20 F.22s in 1950. (below) India was the main customer for the Spitfire T.9 two seat trainer, purchasing 10 conversions.

with Spitfire Vs, VIIIs and IXs, of which three remained in the UK/Europe and the other (No 328) served in the Middle East before operating in Europe. These squadrons also become part of the *Armée de l'Air* in late 1945.

Post war, France received 172 ex RAF Spitfire IXs from early 1947 along with 50 Mk.Vs. Many of the Mk.IXs were sent to French Indo-China (where local unrest was quickly developing) and North Africa, while

the Mk.Vs were used mainly for training. The *Aéronavale* also received 48 Seafire IIs and some Seafire XVs in 1946-1948 and operated 20 of the Spitfire IXs for the training of Seafire pilots.

Greece

Two Greek squadrons (Nos 335 and 336) equipped with Spitfire Vs flew as part of the RAF between from 1943, operating in the Middle East. At the end of the war they were integrated into the Royal Hellenic Air Force which operated 106 Mk.Vs, the last of them not retired until 1951.

Between 1947 and 1949 Greece received 74 Spitfire LF.IXs, three HF.IXs, one PR.XI and 66 LF.XVIs, many of the Mk.IXs from RAF stocks already based in the country. The Greek Spitfires saw active service during the civil war of 1948-49 and the last Mk.XVIs were retired in 1955. Royal Hellenic Air Force Spitfires retained their original RAF serial numbers.

India

The (Royal) Indian Air Force began operating Spitfires alongside RAF units in mid 1944, mainly as fighter-bombers harassing Japanese

Many of France's 172 Spitfire IXs received from 1947 served in French Indo-China, although the aircraft did not suit operations in the very humid and corrosive climate. This lineup comprises aircraft from GC.1/3 'Navarre' and GC.2/3 'Champagne' at Nha Trang in 1950.

Israel's first of about 90 Spitfire IXs inadvertently came from Egypt, the aircraft forced down during a raid on the day of the State of Israel's formation in May 1948. The LF.IX was rebuilt and flown by 101 Squadron until burnt out in 1951.

military units as they withdrew from the country and in Burma. Four squadrons were formed before the end of the war, Nos 4, 8, 9 and 10. The first two squadrons were initially equipped with Spitfire VCs before switching to Mk.VIIIs, these also equipping the other two units.

Overall, India operated about 150 Spitfire VIIIs plus approximately 70 Mk.XIV/XVIIIs and 10 T.9 trainers. By the time of the partition in 1947 only one Spitfire squadron (No 2) remained operational, but tensions in the Kashmir region with newly created Pakistan saw former RAF aircraft which had been left in their crates unpacked, assembled and put into service, again largely as fighter-bombers. The trainers remained in service until the mid 1950s.

Ireland (Eire)

The Irish Air Corps took delivery of 12 'denavalised' Seafire L.IIIs (in effect making them into Spitfire VCs) in 1947 and six Spitfire T.9 trainers in June 1951. The Seafires replaced the Hurricane IIs of No 2 Squadron.

All had seen previous service with Britain, the trainers – like all of their ilk – converted from Spitfire IX fighters. The last T.9s were not withdrawn from service until 1961, leaving Ireland as the world's last military operator of the Spitfire/Seafire line.

Israel

The fledgling Israeli Air Force (*Heyl Ha'Avir*) accidentally received its first Spitfire on the day the State of

Israel was established, 15 May 1948. The date also marked the start of the first Arab-Israeli War as Israel fought for a tenuous survival.

On that day, two Royal Egyptian Air Force Spitfire LF.IXs attacked Tel Aviv's Dov airfield. One was hit by ground fire and came down on a beach. Damaged but salvageable, the Spitfire was repaired and put into Israeli service. The next two taken on strength were built up from parts of scrapped RAF Spitfire IXs.

Israel built up its air force with aircraft often surreptitiously obtained from various sources including from Czechoslovakia which sold Israel a quantity of Avia S.199s, licence built Messerschmitt Bf 109Gs with Junkers Jumo engine.

Czechoslovakia was also the major source of Israeli Spitfires, the first of eventually 56 Mk.IXs from that country arriving in late September 1948. Some were flown to Israel but most were shipped, the aircraft equipping No 101 Squadron which put its Spitfires to considerable combat use. One famous action on 7 January 1949 saw Israeli Spitfire IXs and RAF Spitfire FR.18s clash, with the Israeli aircraft emerging victorious (see *Spitfire Versus Spitfire* breakout box).

Israel signed an agreement for the purchase of 30 Spitfire IXs from Italy in 1951 following the lifting of a United Nations embargo, but they were not delivered for two years, equipping No 105 Squadron. No 101 Squadron had meanwhile re-equipped with P-51 Mustangs and its Spitfires transferred to the newly formed 107 Squadron, although this was disbanded in 1954. No 105 Squadron therefore became the sole Israeli Spitfire unit until it was also disbanded in February 1956.

The Spitfire ended its Israeli career two months later. Thirty aircraft had been refurbished and sold to Burma the previous year.

Italy

Following Italy's surrender in September 1943, the pro Allies Co-Belligerent Air Force acquired 53 Spitfire Vs from RAF stocks of which 33 were made airworthy and put into service with the 20° *Gruppo*, 51° *Stormo* The unit became operational in October 1944 and flew from bases in Italy until the end of the war.

When the Allies withdrew from Italy large numbers of aircraft including

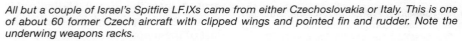

All but a couple of Israel's Spitfire LF.IXs came from either Czechoslovakia or Italy. This is one of about 60 former Czech aircraft with clipped wings and pointed fin and rudder. Note the underwing weapons racks.

Spitfire IXs were left behind, of which about 145 were taken on *Aeronautica Militare Italiana* (AMI) strength from mid 1947. At the time, Italy was limited by treaty to an air force with no more than 200 fighters.

Thirty Spitfire IXs were supplied to Israel in 1953 and the last was retired from Italian service in April 1956.

Netherlands

Two Dutch manned fighter squadrons equipped with Spitfires were established within the RAF in 1943-44, No 318 flying Spitfire Vs and IXs in the Middle East and Italy and No 322 with Mks.V, IX, XVI and XIV in the UK and Europe.

Post war, the *Luchstridkrachten* obtained 55 Spitfire IXs plus Mk.XVIs and four PR.XIs from RAF stocks plus three T.9 trainer conversions. Deliveries began in 1947 and ended the following year. Some of the Spitfires saw combat in 1947-49, the aircraft of 322 Squadron fighting nationalist insurgents in the Netherlands East Indies during the war which resulted in the establishment of independent Indonesia.

Fifteen aircraft were transferred to Belgium in 1950 and the last Spitfires were withdrawn from Dutch service in 1954.

(above) A former Italian Spitfire LF.IX in Israeli service, identifiable by its standard span wings and early shape fin and rudder. (below) Italy received Spitfire Vs for its Co-Belligerent Air Force following the surrender in 1943 and Mk.IXs in 1947. This Mk.VC was one of the original batch and the first sortie was flown in October 1944.

The Netherlands acquired 55 Spitfire IXs and other marks in 1947-48.

(for 331 and 332 Squadrons) and three PR.XIs. These in combination with the 24 Spitfire IXs inherited in 1945 brought the Norwegian total to 74 aircraft.

Both squadrons re-equipped with Republic F-84G Thunderjets in 1952-53 and the last Spitfire was retired in 1954.

Portugal

Portugal was an early potential customer for the Spitfire, ordering 15 Mk.Is in 1939 for delivery by March 1940. Events obviously intervened but in February 1940 the Air Ministry agreed to divert 12 Mk.Is from the RAF to Britain's oldest ally, but once again events (this time Germany's invasion of France and the Low Countries) prevented the agreement being consummated.

Deliveries of Spitfires to Portugal finally began in November 1942 when 18 surplus Mk.Is were sent. In August 1943 the two countries completed the Azores Agreement under which Portugal would allow Britain access to shipping facilities in the Azores in exchange for 33 ex RAF Spitfire VBs. These were delivered from October 1943.

More Spitfires were delivered to the *Força Aérea Portuguesa* after the war with 60 Mk.Vs handed over in 1947-48. The aircraft were withdrawn from front line service in 1951 and replaced by Republic F-84G Thunderjets.

South Africa

Unlike the other Commonwealth countries, South Africa's substantial contribution to the RAF's war effort was not through newly formed squadrons within the RAF system but via the SAAF's own squadrons. Nos 1, 2, 3, 4, 7, 9, 10, 11, 40 and 41 Squadrons operated as part of the Desert Air Force in North Africa and Italy, between them flying mainly Spitfire Vs and IXs. No 1 Squadron also had Spitfire VIIIs and tactical reconnaissance specialist 40 Squadron some PR.XIs.

Nos 1, 2 and 4 Squadrons were reformed in South Africa after the war, equipped with 136 Spitfire IXs donated by the British Government.

(above) Some Dutch Spitfires saw action during the war of independence which resulted in the Netherlands East Indies becoming Indonesia in 1949. (below) The three Spitfire T.9 trainers acquired by the Netherlands lined up at Eastleigh in March 1948 before delivery.

New Zealand

Of the several Royal New Zealand Air Force squadrons established in Britain, only one (No 485) was equipped with Spitfires. Established in April 1941, it was the first Commonwealth squadron to be formed in the United Kingdom. It flew Spitfire Mks.I, II, V, IX and XVIs between then and 1945, operating from European bases after the invasion.

Norway

The two Norwegian Spitfire squadrons which had operated under RAF control from 1941-42 (Nos 331 and 332) became part of the Royal Norwegian Air Force in November 1945, at the time equipped with Spitfire IXs.

Norway began purchasing ex RAF Spitfires in 1947 as part of a general defence re-equipment programme, the package including 47 Spitfire IXs

The Royal Norwegian Air Force purchased 74 ex RAF Spitfires from 1947, among them three PR.XIs. They were operated by No 1 Photographic Reconnaissance Wing and are shown here at their base at Gardermoen.

No 60 Squadron also had Spitfires. South Africa's Spitfires began to be replaced in 1950, No 2 Squadron swapping them for P-51 Mustangs before going to fight in the Korean War, while de Havilland Vampires replaced the others from 1952.

Soviet Union

The first Soviet request for Spitfires came in October 1942 via the Soviet Ambassador in London. Winston Churchill approved the request and the first aircraft (143 Mk.VBs) were handed over in March 1943, these taken from the RAF's Middle East stocks. Another 1,186 LF.IXs and two HF.IXs followed in 1944-45, bringing the total to 1,331.

The Soviet Spitfires fought mainly in the north including with great effect during the defence of Stalingrad. Some were locally converted to dual seat trainers with a second cockpit located behind the original.

Typical of the Soviets, it has been noted that for political reasons, the Spitfire (or any other aircraft supplied by the West in vast numbers and never paid for) is not mentioned in the official histories of the Soviet Air Force in World War II.

Of the 1,331 Spitfires obtained by the Soviet Union under Lend-Lease, the first 143 were Mk.VBs, most of them well used examples from the RAF's Middle East stocks. These examples are ready for handover and have their red stars painted on but with British serials retained. BM185 is in the foreground.

Sweden

The Swedish Government maintained its policy of neutrality throughout the war despite tempting offers from both sides. Britain offered to provide 200 Spitfires if the supply of ball bearings to Germany was stopped, but to no avail.

Faced with a perceived threat from the Soviet Union, Sweden began a military modernisation programme in 1948 which included the delivery of 50 Spitfire PR.XIXs from

Reflections on a Swedish Spitfire PR.19, one of 50 delivered from RAF stocks between 1948 and 1955. This particular aircraft is the former PS875 and carries the Swedish serial 31040 and tail number '60'.

Swedish Spitfire PR.19s being prepared for delivery in October 1948. The Swedish designation for the aircraft was S.31.

RAF stocks between then and 1955. In Swedish service the Spitfires were designated the S.31 and used for reconnaissance flights over the Gulf of Bothnia and the Baltic Sea.

Syria

Syria received 10 Spitfire F.22s from surplus RAF stocks in 1950, delivered after refurbishment by Airwork Ltd. They remained in service until 1953 when they were replaced by Gloster Meteor F.8s.

Thailand

Thirty reconditioned Spitfire FR.XIVs were delivered to Thailand from 1948, the aircraft replacing a variety of fighters which had remained in the country at the conclusion of the war. These included some Japanese types. The Spitfires were themselves replaced by Grumman F8F Bearcats from the mid 1950s.

Turkey

Turkey remained officially neutral throughout most of the war, not declaring war on Germany until March 1945 when the battle was as good as over. The *Türk Hava Kuvvetleri* was one of the air forces which had sought to purchase Spitfires before 1939, requesting up to 60 aircraft. Despite the difficulties and due to Britain's desire to keep Turkey 'on side', two Mk.Is were supplied in May 1940 but although being shipped it is doubtful they were delivered.

Efforts to keep in Turkey's good books continued, with the result that some 84 tropicalised Spitfire VBs and VCs were supplied in 1944. Germany was equally keen to keep the Turks happy, supplying a number of Focke-Wulf 190s and creating the strange sight of the two foes operating side by side!

Post war, substantial amounts of British surplus military equipment was received including 203 Spitfire IXs and four PR.XIXs. The last was retired in 1954.

A Royal Thai Air Force Spitfire FR.14E, one of 30 refurbished examples delivered from 1948.

One of the two Spitfire Is allocated to Turkey in 1940 and photographed here (possibly in Italy during 1943) in RAF hands. There is doubt as to whether they were ever formally delivered to Turkey despite being shipped. This aircraft is the former P9567 and the serial 'HK856' is not one applied to any Spitfire but from a block reserved for impressed aircraft. The other aircraft was P9566/HK854. Turkey subsequently received Spitfire Vs in 1944 plus IXs and XIXs post war. (via Neil Mackenzie)

United States of America

Although no Spitfires were purchased directly by the USAAF, about 600 were operated under a 'Reverse Lend-Lease' arrangement after the USA's entry into the war in December 1941. Before that, three American Eagle Squadrons had been formed within RAF Fighter Command, manned by pilots who wished to help Britain in the fight against Nazi Germany (see chronology entry for 19 October 1940).

These were disbanded on 29 September 1942 and transferred to the 4th Fighter Group of the 8th Air Force at Debden as the 334th, 335th and 336th Pursuit Squadrons. In June 1942 the 31st and 52nd Fighter Groups arrived in England and were equipped with Spitfire Vs.

Both Groups were transported to Gibraltar under the command of the 12th AF in late October, their previous aircraft allocated to the 67th Observation Group at Membury, Wiltshire, for

the training of reconnaissance pilots. The 7th PR Group at Mount Farm (Oxfordshire) was also equipped with Spitfire Vs on its formation in mid 1943 before switching to PR.XIs. The Spitfire PR.XI also equipped the 10th, 67th and 363rd Tactical Reconnaissance Groups of the 9th AF, by then operating from England as part of the tactical air force assembling for the invasion of Europe.

Meanwhile, the 31st and 52nd FGs had participated in the Operation

Spitfire VB EN851 'Lima Challenger' of the USAAF's 307th Fighter Squadron, 31st Fighter Group, based in England in 1942. This was the first 'proper' USAAF Spitfire unit in Britain, taking into account the American 'Eagle' squadrons previously formed within the RAF structure.

The USAAF operated some 600 Spitfires during the war, most of them Mk.Vs and PR.XIs. These Mk.VBs are from the 12th and 109th Reconnaissance Squadrons, 67th Reconnaissance Group, operating out of Membury in England during 1943.

British based Spitfire PR.XIs PA950 (top) and PA944 of the USAAF's 7th Photo Group. The unit's base was at Mount Farm near Oxford.

(below) Spitfire VC JK226 of the USAAF's 308th FS, 31st FG, photographed in the Mediterranean area in late 1942.

Torch landings in North Africa in November 1942, operating from Algiers. After strafing German troop and vehicle columns in the desert they were used on ground attack and bomber escort missions before participating in the July 1943 invasion of Sicily. From April 1944 they began to re-equip with the P-51 Mustang and were transferred to the 15th AF in Italy.

Some other 8th Air Force units in England had one or two Spitfires on strength, including the Fighter Training and HQ Squadrons. Spitfire VC AA963 was despatched to the USA in February 1942 for evaluation and exhibition purposes.

Yugoslavia

One Yugoslav squadron (No 352) was established within the RAF's Balkan Air Force in 1944 equipped with Spitfire VCs. They were used on ground attack and bomber escort missions and carried Yugoslav insignia (a red star over the RAF roundel) in combination with British serial numbers.

The squadron was disbanded by the RAF in June 1945 and became part of the *Jugoslovensko Ratno Vazduhoplovstvo*, the Spitfires remaining in service until 1952 when the last of 18 Mk.VCs and three Mk.IXs was retired.

SPITFIRE and SEAFIRE SERIAL NUMBERS

Note: British military serial numbers have long followed an alphabetical/numerical sequence, as reflected below. Most wartime serial blocks had 'blacked out' or unused sections with some numbers missing for security purposes. It was therefore possible for a serial block to cover, for example 1,000 numbers but contain only, say, 700 actual serials/aircraft due to these 'blackout blocks'. Order cancellations and amendments also left some gaps within the overall serial range from time to time.

The following summarises the Spitfire and Seafire serial blocks, build dates, the serial numbers actually used and other points of interest. Unless otherwise stated, aircraft were built by Supermarine.

SPITFIRE

K5054
Prototype Supermarine Type 300 ordered in January 1935 to Air Ministry specification F.37/34 and first flown 5 March 1936.

K9787-L1096
First order for 310 Spitfire Is placed in June 1936, delivered between July 1938 and September 1939. First 74 with Merlin II engine, remainder with Merlin III. One additional aircraft for France delivered July 1939. K9834 completed as Type 323 Speed Spitfire for attempt on world's air speed record, first flown January 1939 with Class B registration N.17.
Serial numbers: K9787-9999 and L1000-1096.

N3023-N3299
200 Spitfire Is ordered December 1938 and delivered August 1939-January 1940. N3297 completed as Spitfire III prototype May 1940.
Serial numbers: N3023-3072, N3091-3130, N3160-3203, N3221-3250 and N3264-3299.

P7280-P8799
1,000 ordered April 1939 as first to be built at new Castle Bromwich Aircraft factory in Birmingham. Built as 751 Mk.IIA, 170 Mk.IIB and 79 Mk.VA/B between June 1940 and July 1941.
Serial numbers: P7280-7329, P7350-7389, P7420-7449, P7490-7509, P7520-7569, P7590-7629, P7661-7699, P7730-7759, P7810-7859, P7880-7929, P7960-7999, P8010-8049, P8070-8099, P8130-8149, P8160-8209, P8230-8279, P8310-8349, P8360-8399, P8420-8449,

Spitfire II P8078.

P8460-8479, P8500-8549, P8560-8609, P8640-8679, P8690-8679 (all Mk.II to here), P8690-8729, P8740-8759 and P8780-8799.
Conversions: 47 Mk.II to Mk.V; P7505 and P8784 to PR.XIII.

Spitfire IIB P8332.

P9305-P9584
200 Spitfire Is ordered in April 1939 and of which 181 delivered to the RAF from January 1940. Two additional aircraft (P9566-9567) delivered to Turkey in May 1940 and 17 aircraft (P9568-9584) cancelled.
Serial numbers: P9305-9339, P9360-9399, P9420-9469, P9490-9519 and P9540-9565.
Conversions: P9307-9309, P9315, P9384, P9551-9552 to photo-reconnaissance; Mk.IIs P9367, P9448, P9540, P9550, P9556 and P9563 to Mk.VA.

P9551 and P9552
Two Spitfire PR.IV (Type D) prototypes ordered May 1940 and delivered September 1940 and February 1941.

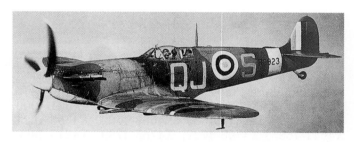

Spitfire VB R6923 converted from Mk.I.

R6595-R7350
450 Spitfire Is ordered August 1939 and built as 343 F.I (of which 43 converted to Mk.V), 65 F.VA, 24 F.VB, 6 PR Type C and 12 PR.IV between April 1940 and April 1941.
Serial numbers: R6595-6644 (Mk.I), R6683-6722 (Mk.I), R6751-6780 (Mk.I), R6799-6818 (Mk.I), R6829-6840 (Mk.I), R6879-6928 (Mk.I), R6957-6996 (Mk.I), R7015-7028 (Mk.I), R7029-7034 (PR Type C), R7035-7044 (PR.IV), R7055-7056 (PR.IV), R7057-7074 (Mk.I), R7114-7163 (Mk.I), R7192-7218 (Mk.I), R7219-7231 (Mk.V), R7232-7249 (Mk.I), R725-7252 (Mk.I), R7253-7256 (Mk.V), R7257 (Mk.I), R7258-7279 (Mk.V), R7290-7309 (Mk.V) and R7333-7350 (Mk.V).
Conversions: R6684 and R6924 to Mk.II; R6602, 6620, 6720, 6722, 6759, 6761, 6770, 6776, 6801, 6809, 6817, 6882, 6888-6890, 6897, 6908, 6911-6913, 6919, 6923-6924, 6957, 6960, 6992, 7022, 7060, 7127, 7158, 7161, 7192, 7194-7196, 7205, 7207-7210, 7213 and 7217-7218 Mk.I to Mk.V; R7130 and R7308 to PR.VII; R7335 to PR.XIII.

Spitfire VB W3368.

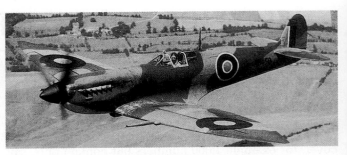

Spitfire VII AB450.

W3109-W3970

450 Spitfire Is ordered March 1940 and built as 25 Mk.VA and 425 Mk.VB between April and October 1941.

Serial numbers: W3109-3138, W3168-3187, W3207-3216, W3226-3265, W3305-3334, W3364-3383, W3403-3412, W3422-3461, W3501-3530, W3560-3579, W3599-3608, W3618-3657, W3697-3726, W3756-3775, W3795-3804, W3814-3853, W3893-3902 and W3931-3970.

Conversions: W3237 to Mk.III; W3112, W3135, W3831 to PR.XIII; W3760 to floatplane. Also three conversions to Seafire IB with new serial numbers – W3212 to NX883, W3371 to PA119, W3372 to NX980.

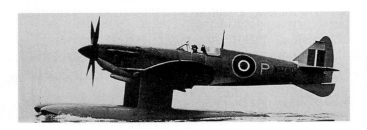

Spitfire V Floatplane W3760.

X4009-X4997

500 Spitfire Is ordered in June 1940 and built as 484 Mk.I, 8 PR.IV, 4 Mk.VA and 4 Mk.VB July 1940-January 1941.

Serial numbers: X4009-4038, X4051-4070, X4101-4110, X4159-4188, X4231-4280, X4317-4356, X4381-4390, X4409-4428, X4471-4505, X4538-4562, X4585-4624, X4641-4685, X4708-4722, X4765-4789, X4815-4859, X4896-4945 and X4988-4997.

Conversions: to Mk.II – X4067, X4622, X4776; to Mk.V – X4062, X4106, X4257-4258, X4272, X4280, X4555, X4604-4606, X4620-4624, X4663-4671, X4902, X4922, X4997; to Mk.VI – X4942; to PR.XIII (1943) – X4021, X4615, X4660, X4766; to Seafire IB – X4717, X4989. Others also converted to PR models.

Spitfire PR Type F X4492.

AA718-AB536

500 Spitfire Is ordered in July 1940 and built as 78 PR.IV, 405 Mk.V, 12 Mk.VI and 5 Mk.IX August 1941-February 1942.

Serial numbers: AA718-767 (Mk.VB), AA781-815 (PR.IV), AA833-873 (Mk.VB), AA874 (Mk.VC), AA875-882 (Mk.VB), AA902-946 (Mk.VB), AA963 Mk.VC), AA964-967 (Mk.VB), AA968 (Mk.VC), AA969-975 (Mk.VB), AA976-977 (Mk.VC), AA978-979 (Mk.VB), AA980 (Mk.VC), AA981-982 (Mk.VB), AB118-123 (PR.IV), AB124 (Mk.VB), AB125-129 (PR.IV), AB130 (Mk.VA), AB131-132 (PR.IV), AB133-152 (Mks.VB/C), AB167-216 (Mk.VB/C), AB240-254 (Mk.VB/C), AB255-284 (Mk.VB), AB300-319 (PR.IV), AB320-349 (Mk.VB), AB363-364 (Mk.VB), AB365 (Mk.VC), AB366 (Mk.VB), AB367-368 (Mk.VC), AB369-370 (Mk.VB), AB371-372 (Mk.VC), AB373 (Mk.VB), AB374 (Mk.VC), AB375-376 (Mk.VB), AB377 (Mk.VC), AB378-379 (Mk.VB), AB380-381 (Mk.VC), AB382 (Mk.VB), AB401-416 (Mk.VB), AB417 (Mk.VC), AB418-420 (Mk.VB), AB421-430 (PR.IV), AB450 (Mk.VC), AB451 (Mk.VB), AB452-453 (Mk.VC), AB454 (Mk.VB), AB455 (Mk.VC), AB456-460 (Mk.IX), AB461-469 (Mk.VC), AB487 Mk.VB), AB488-489 (Mk.VC), AB490 (Mk.VB), AB491-497 (Mk.VC), AB498 (Mk.VI), AB499 (Mk.VC), AB500 (Mk.VB), AB501 (Mk.VC), AB502 (Mk.VB), AB503 (Mk.VI), AB504-505 (Mk.VC), AB506 (Mk.VI), AB507-512 (Mk.VC), AB513 (Mk.VI), AB514-515 (Mk.VC), AB516 (Mk.VI), AB517-522 (Mk.VC), AB523 (Mk.VI), AB524-526 (Mk.VB/C), AB527-530 (Mk.VI), AB531-532 (Mk.VB/C), AB533-534 (Mk.VI) and AB535-536 (Mk.VC).

Conversions: AA739 to PR.XIII; AB176, 200, 211 to Mk.VI; AB499 to floatplane; AB450 to prototype Mk.VII; others to Seafire IB with new serials.

AB779-AD584

500 Spitfire Is ordered from Castle Bromwich in June 1940 but all built as F.VBs between July and November 1941.

Serial numbers: AB779-828, AB841-875, AB892-941, AB960-994, AD111-140, AD176-210, AD225-274, AD288-332, AD348-397, AD411-430, AD449-478, AD498-517 and AD535-584.

Conversions: AD 534 and AD501 to PR.XIII; AD366 to Mk.IX; several to Seafire IB with new serials.

AR212-AR621

300 Spitfire Is ordered from Westland Aircraft in August 1940 and built as 50 F.IA, 140 F.VB and 110 F.VC July 1941-September 1942.

Serial numbers: AR212-261 (F.IA), AR274-298 (F.VB), AR318-347 (F.VB), AR362-406 (F.VB), AR422-461 (F.VB), AR462-471 (F.VC), AR488-532 (F.VC), AR546-570 (F.VC) and AR592-621 (F.VC).

Conversions: AR319 to PR.XIII; AR384, AR442-448, AR457-461 to Seafire IB direct from production; several F.IAs to PR.IV and Mk.V.

Spitfire VB BL450.

BL231-BM653

1,000 Spitfire IIIs ordered from Castle Bromwich in October 1940 but built as F.VBs between November 1941 and May 1942.

Serial numbers: BL231-267, BL285-304, BL311-356, BL365-391, BL403-450, BL461-500, BL509-551, BL562-600, BL613-647, BL655-699, BL707-736, BL748-789, BL801-833, BL846-864, BL887-909, BL918-941, BL956-998.

BM113-162, BM176-211, BM227-274, BM289-329, BM343-386, BM402-430, BM447-493, BM508-543, BM556-597 and BM624-653.

Conversions: BL526, BM447 and BM591 to PR.XIII; others to Seafire IB.

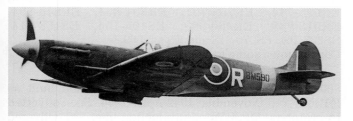

Spitfire VB BM590.

BP844-BS152

800 Spitfire Is ordered October 1940 of which 608 built as 104 PR.IV, 11 F.VB, 349 F.VC, 69 F.VI, 2 HF.VII and 73 F.IXC between February and November 1942.

Serial numbers: BP844-854 (F.VB), BP855-878 (F.VC), BP879-892 (PR.IV), BP904-937 (PR.IV), BP950-993 (F.VC).

BR106-137 (F.VC), BR138 (F.IX), BR139 (F.VC), BR140-143 (F.IX), BR159 (F.VI), BR160-161 (F.VC), BR162 (F.VI), BR163 (F.VC), BR164 (F.VI), BR165-166 (F.VC), BR167 (F.VI), BR168-170 (F.VC), BR171-172 (F.VI), BR173 (F.VC), BR174 (F.VI), BR175-177 (F.VC), BR178 (F.VI), BR179-180 (F.VC), BR181 (F.VI), BR182-185 (F.VC), BR186 (F.VI), BR187-188 (F.VC), BR189 (F.VI), BR190 (F.VC), BR191 (F.VI), BR192 (F.VC), BR193 (F.VI), BR194-196 (F.VC), BR197 (F.VI), BR198-199 (F.VC), BR200 (F.VI), BR201-204 (F.VC), BR205 (F.VI), BR226-242 (F.VC), BR243 (F.VI), BR244-246 (F.VC), BR247 (F.VI),

Spitfire VC BR539.

BR248-249 (F.VC), BR250 (F.VI), BR251 (F.VC), BR252 (F.VI), BR253-254 (F.VC), BR255 (F.VI), BR256 (F.VC), BR282-285 (F.VC), BR286-287 (F.VI), BR288 (F.VC), BR289 (F.VI), BR290-296 (F.VC), BR297-298 (F.VI), BR299-301 (F.VC), BR302 (F.VI), BR303-304 (F.VC), BR305 (F.VI), BR306 (F.VC), BR307 (F.VI), BR308 (F.VC), BR309-310 (F.VI), BR311-313 (F.VC), BR314 (F.VI), BR315-317 (F.VC), BR318-319 (F.VI), BR320-325 (F.VC), BR326 (F.VI), BR327-328 (F.VC), BR329-330 (F.VI), BR344-357 (F.VC), BR358 (PR.IV), BR359-360 (F.VC), BR361 (PR.IV), BR362-363 (F.VC), BR364 (PR.IV), BR365-368 (F.VC), BR369-371 (F.IXC), BR372-393 (F.VC), BR410-435 (PR.IV), BR459-499 (F.VC), BR515-549 (F.VC), BR562 (F.VC), BR563 (F.VI), BR564-566 (F.VC), BR567 (F.VI), BR568-570 (F.VC), BR571 (F.VI), BR572-574 (F.VC), BR575 (F.VI), BR576 (F.VC), BR577-579 (F.VI), BR580 (F.VC), BR581 (F.IXC), BR582-584 (F.VC), BR585 (F.VI), BR586 (F.VC), BR587-588 (F.VI), BR589 (F.VC), BR590 (F.VI), BR591 (F.VC), BR592 (F.IXC), BR593 (F.VI), BR594 (F.IXC), BR595 (F.VI), BR596 (F.IXC), BR597-599 (F.VI), BR600-605 (F.IXC), BR621-640 (F.VC), BR641-670 (PR.IV), BR977-978 (F.IXC), BR979 (F.VI), BR980-982 (F.IXC), BR983-984 (F.VI), BR985-986 (F.IXC), BR987 (F.VI).

BS104-105 (F.IXC), BS106 (F.VI), BS107 (F.IXC), BS108 (F.VI), BS109-110 (F.IXC), BS111 (F.VI), BS112-113 (F.IXC), BS114-115 (F.VI), BS116 (F.IXC), BS117 (F.VI), BS118-120 (F.IXC), BS121 (HF.VII), BS-122-123 (F.IXC), BS124 (F.VI), BS125-132 (F.IXC), BS133-134 (F.VI), BS135-140 (F.IXC), BS141 (F.VI), BS142 (HF.VII), BS143-145 (F.IXC), BS146 (F.VI), BS147-148 (F.IXC), BS149 (F.VI) and BS150-152 (F.IXC).

Spitfire IXCs BS306 (above) and BS342.

BS157-BS559

300 Spitfire Is ordered in October 1940 and delivered from June 1942 as 25 PR.IV, 122 F.VC, 10 F.VI, 3 HF.VII and 140 F.IX.

Serial numbers: BS157-202, BS218-255, BS271-319, BS335-367, BS383-411, BS427-474, BS489-515 and BS530-559.

PR.IVs were BS355-367, BS489-496, BS500 and BS503-505.

F.VIs were BS228, BS245, BS436-437, BS442, BS448, BS453, BS460, BS465 and BS 472.

HF.VIIs were BS229, BS253 and BS427.
Note: Mk.IX trials their aircraft BS274 carried the incorrect serial BF274 for most of its life, this clashing with the real BF274, a Bristol Blenheim.

DP845 and DP851
Two Griffon powered Spitfire Mk.IV prototypes ordered May 1941. DP845 first flight 27 November 1941, DP851 December 1941. Both renamed as Mk.XX.

EE600-EE867
200 Spitfire Vs ordered from Westland Aircraft in September 1941 and built as F.VCs between September 1942 and February 1943.
Serial numbers: EE600-644, EE657-690, EE713-753, EE766-811 and EE834-867.

EF526-EF753
185 Spitfire Vs ordered from Westland Aircraft in October 1941 as built as F.VCs between February and November 1943. EF753 was last Westland built Spitfire but company continued manufacturing Seafires.
Serial numbers: EF526-570, EF584-616, EF629-656, EF671-710 and EF715-753.

Spitfire IXC EN133

EN112-EN759
500 Spitfire VCs ordered in August 1941 of which 440 built as 7 PR.IV, 19 HF.VII, 288 F/LF.IX, 81 PR.XI and 45 F.XII. Remainder cancelled.
Serial numbers: EN112-152 (F.IX), EN153 (PR.IV), EN154 (F.IX), EN155 (PR.IV), EN156 (F.IX), EN171-177 (F.IX), EN178 (HF.VII), EN179-191 (F.IX), EN192 (HF.VII), EN193-207 (F.IX), EN221-238 (F.XII), EN239-261 (F.IX), EN262 (PR.IV), EN263 (F.IX), EN264 (PR.IV), EN265-270 (F.IX), EN285 (HF.VII), EN286-296 (F.IX), EN297 (HF.VII), EN298-309 (F.IX), EN310 (HF.VII), EN311-315 (F.IX), EN329-337 (F.IX), EN338 (PR.XI), EN339-340 (F.IX), EN341-343 (PR.XI), EN344-345 (F.IX), EN346-348 (PR.XI), EN349-370 (F.IX), EN385 (PR.XI), EN386-390 (F.IX), EN391 (PR.XI), EN392-394 (F.IX), EN395-396 (PR.XI), EN397-406 (F.IX), EN407-430 (PR.XI), EN444-456 (F.IX), EN457 (HF.VII), EN458-464 (F.IX), EN465 (HF.VII), EN466-469 (F.IX), EN470 (HF.VII), EN471-473 (F.IX), EN474 (HF.VII), EN475-476 (F.IX), EN477 (HF.VII), EN478-483 (F.IX), EN490-493 (F.IX), EN494-497 (HF.VII), EN498 (F.IX), EN499 (HF.VII), EN500-502 (F.IX), EN503-504 (PR.XI), EN505-506 (HF.VII), EN507-508 (PR.XI), EN509 (HF.VII), EN510 (F.IX), EN511-512 (HF.VII), EN513-534 (F.IX), EN551-571 (F.IX), EN572-583 (LF.IX), EN601-627 (F.XII), EN628-637 (LF.IX) and EN652-685 (PR.XI). Following serials cancelled.

Spitfire V EP455.

EN763-ER200
905 Spitfire Vs ordered from Castle Bromwich in August 1941 and built as Mk.VB and Mk.VC between April and August 1942. EN830 captured by Germans and fitted with Daimler-Benz DB 605A engine for testing.
Serial numbers: EN763-800, EN821-867, EN887-932, EN944-98. P107-152, EP164-213, EP226-260, EP275-316, EP327-366, EP380-417, EP431-473, EP485-523, EP536-579, EP594-624, EP636-669, EP682-729, EP747-795, EP812-847, EP869-915, EP951-990, ER114-146 and ER159-200.
Conversions: some to Seafire IB.

ER206-ES369
750 Spitfire IVs ordered from Castle Bromwich in August 1941 but built as Mk.VB and Mk.VC August-December 1942. Most delivered to North Africa and Middle East.
Serial numbers: ER206-229, ER245-283, ER299-345, ER461-510, ER524-571, ER583-626, ER634-679, ER695-744, ER758-791, ER804-834, ER846-894, ER913-948, ER960-998, ES105-154, ES168-214, ES227-264, ES276-318 and ES335-369.
Conversions: some to Mk.IX.

JF274-JG695
800 Spitfire VIIIs ordered January 1942 and built as F/HF/LF.VIII November 1942-April 1944. Six converted to prototype Mk.XIVs on production line, 135 delivered to RAAF.
Serial numbers: JF274-300 (F.VIII), JF316-321 (F.XIV prototypes), JF322-327 (F.VIII), JF328 (HF.VIII), JF329-364 (F.VIII), JF392-427 (F.VIII), JF443-461 (F.VIII), JF462 (LF.VIII), JF463-485 (F.VIII), JF501-502 (F.VIII), JF503-504 (LF.VIII), JF505-513 (F.VIII), JF514-515 (LF.VIII), JF516-528 (F.VIII), JF557-592 (F.VIII), JF613-630 (F.VIII), JF658-676 (F.VIII), JF692-716 (F.VIII), JF740-789 (LF.VIII), JF805-850 (LF.VIII), JF869-893 (LF.VIII), JF894-899 (F.VIII), JF900-902 (LF.VIII), JF926-967 (LF.VIII).
JG104-124 (LF.VIII), JG157-159 (LF.VIII), JG160 (F.VIII), JG161 (LF.VIII), JG162 (F.VIII), JG163-165 (LF.VIII), JG166 (F.VIII), JG167-204 (LF.VIII), JG239-275 (LF.VIII),

Spitfire LF.VIII JG655.

JG312-356 (LF.VIII), JG371-387 (LF.VIII), JG404-432 (LF.VIII), JG465-500 (LF.VIII), JG527-568 (LF.VIII), JG603-624 (LF.VIII) and JG646-695 (LF.VIII).
Conversions: JF316-321 to Mk.XIV prototypes; JG204 fitted with laminar flow wing for proposed Mk.23; JF299 first Spitfire fitted with 'bubble' canopy and cut down rear fuselage decking.

JG713-JL395

989 Spitfire Vs ordered from Castle Bromwich in January 1942 of which 913 completed as Mk.VC and 76 converted to F/LF.IX on production line or after manufacture. Aircraft built between December 1942 and April 1943 and the vast majority were delivered Gibralter and North West Africa.
Serial numbers: JG713-752, JG769-810, JG835-852, JG864-899, JG912-960, JK101-145, JK159-195, JK214-236, JK249-285, JK303-346, JK359-408, JK425-472, JK506-551, JK600-620, JK637-678, JK705-742, JK756-796, JK803-842, JK860-892, JK922-950, JK967-992, JL104-140, JL159-188, JL208-256, JL301-338 and JL346-395.
Converted to F.IX: JG722, JG739, JK429, JK611, JK620, JK641, JK659, JK668, JK762, JK769-770, JK795-796, JK860, JK880-884, JK979-980, JL134-138, JL217, JL223, JL226-230, JL239, JL252-256, JL375-377 and JL383-384.
Converted to LF.IX: JK535, JK840, JL106-111, JL159, JL163, JL165, JL172, JL177-180, JL347, JL349, JL351, JL353-354, JL356, JL369, JL361, JL364, JL366, JL369-70, JL372-373, JL385 and JL395.

Spitfire F.21 LA188

LA187-LA332

120 Spitfire 21s ordered from Castle Bromwich in 1943 and built between July 1944 and December 1945, final delivery January 1946. This order was much changed and originally covered 300 Spitfire VCs with serials in the range LA187-582. Later numbers were subsequently allocated to a Seafire 45/46 order (refer Seafire section).
Serial numbers: LA187-236, LA249-284 and LA299-332.
Note: LA188 subsequently modified for compressibility trials, reaching Mach 0.89 in dive.

LV643-LV756

70 photo-reconnaissance Spitfires ordered April 1942 but built as LF.VIIIs between November 1943 and January 1944. Most shipped to India.
Serial numbers: LV643-681, LV726-756.

LZ807-MA906

680 Spitfire VCs ordered from Castle Bromwich in May 1942 and built as 279 Mk.VC and 401 Mk.IX between March and June 1943, final delivery July 1943.
Serial numbers: LZ807-815 (F.VC), LZ816 (LF.IXB), LZ817-830 (F.VC), LZ831-833 (F.IX), LZ834-835 (F.VC), LZ836-843 (F.IX), LZ844-848 (F.VC), LZ861-899 (F.VC), LZ915-925 (F.IX), LZ926-956 (F.VC), LZ969-988 (F.IX).

Spitfire VC MA359.

MA221-260 (F.IX), MA261-266 (F.VC), MA279-297 (F.VC), MA298 (LF.IX), MA299-315 (F.VC), MA328-368 (F.VC), MA369 (F.IX), MA383-397 (F.VC), MA398-428 (F.IX), MA443-487 (F.IX), MA501-546 (F.IX), MA559-601 (F.IX), MA615-643 (F.IX), MA644-657 (F.VC), MA670-704 (F.VC), MA705-713 (F.IX), MA726-767 (F.IX), MA790-819 (F.IX), MA831-849 (F.IX), MA850-863 (F.VC), MA877 (F.VC), MA878-879 (F.IX) and MA880-906 (F.VC).
F.VC to Mk.IX conversions: MA329, MA357, MA645-646, MA648, MA651, MA655, MA657, MA687, MA690, MA860.

MB761-MD403

426 Spitfire VCs ordered in May 1942 and built as 126 F.VII, 163 LF.VIII, 1 F.IX, 10 PR.X, 71 PR.XI and 55 F.XII between July 1943 and March 1944.
Serial numbers: MB761-769 (F.VII), MB770-793 (PR.XI), MB794-805 (F.XII), MB806 (F.VII), MB807 (F.IX), MB808 (F.VII), MB820-828 (F.VII), MB829-863 (F.XII), MB875-882 (LF.XII), MB883-887 (F.VII), MB888-911 (PR.XI), MB912-916 (F.VII), MB929-935 (F.VII), MB936-958 (PR.XI), MB959-976 (LF.VIII), MD100-146 (F.VII), MD159-190 (F.VII), MD191-199 (PR.X), MD213 (PR.X), MD314-256 (LF.VIII), MD269-303 (LF.VIII), MD315-356 (LF.VIII) and MD369-403 (LF.VIII).
Note: most LF.VIIIs delivered direct to India and Australia.

MH298-ML428

2,190 Spitfire VCs ordered from Castle Bromwich in May 1942 and built as 2143 Mk.IX and 47 Mk.VC between July 1943 and May 1944.
Serial numbers: MH298-311 (F.VC), MH312-336 (F.IX), MH349-390 (F.IX), MH413-456 (LF.IX), MH470-496 (LF.IX), MH497 (F.IX), MH498-512 (LF.IX), MH526-563 (LF.IX), MH564-568 (F.VC), MH581-596 (F.VC), MH597-599 (F.IX), MH600 (F.VC), MH601-604 (F.IX), MH605 (F.VC), MH606-626 (F.IX), MH635-636 (F.IX), MH637-646 (F.VC), MH647-678 (F.IX), MH691-711 (F.IX), MH712-738 (LF.IX), MH750-800 (LF.IX), MH813-856 (F.IX), MH869-912 (F.IX), MH924-949 (F.IX), MH950-958 (LF.IX), MH970-999 (F.IX).

Spitfire XII MB882.

Spitfire LF.IX MH448.

MJ114-128 (LF.IX), MJ129-130 (F.IX), MJ131-156 (LF.IX), MJ169-203 (LF.IX), MJ215-227 (F.IX), MJ228-258 (LF.IX), MJ271-305 (LF.IX), MJ306-314 (F.IX), MJ328-369 (LF.IX), MJ382-428 (LF.IX), MJ441-485 (LF.IX), MJ498-536 (LF.IX), MJ549-589 (LF.IX), MJ602-646 (LF.IX), MJ659-698 (LF.IX), MJ712-756 (LF.IX), MJ769-801 (LF.IX), MJ814-858 (LF.IX), MJ870-913 (LF.IX), MJ926-967 (LF.IX), MJ979-999 (LF.IX).

MK112-158 (LF.IX), MK171-213 (LF.IX), MK226-268 (LF.IX), MK280-326 (LF.IX), MK339-379 (LF.IX), MK392-428 (LF.IX), MK440-486 (LF.IX), MK499-534 (LF.IX), MK547-590 (LF.IX), MK602-646 (LF.IX), MK659-669 (LF.IX), MK670-699 (LF.IX), MK713-756 (LF.IX), MK769-812 (LF.IX), MK826-868 (LF.IX), MK881-926 (LF.IX), MK939-967 (LF.IX), MK981-999 (LF.IX).

ML112-156 (LF.IX), ML169-216 (LF.IX), ML229-230 (HF.IX), ML231 (LF.IX), ML232-236 (HF.IX), ML237 (LF.IX), ML238 (HF.IX), ML239 (LF.IX), ML240-241 (HF.IX), ML242 (LF.IX), ML243 (HF.IX), ML244 (LF.IX), ML245-247 (HF.IX), ML248 (LF.IX), ML249 (HF.IX), ML250 (LF.IX), ML251 (HF.IX), ML252 (LF.IX), ML253 (HF.IX), ML254 (LF.IX), ML255 (HF.IX), ML256-258 (LF.IX), ML259 (HF.IX), ML260-277 (LF.IX), ML291-295 (LF.IX), ML296 (HF.IX), ML297-323 (LF.IX), ML338 (HF.IX), ML340-342 (LF.IX), ML343 (HF.IX), ML344-347 (LF.IX), ML348 (HF.IX), ML349-352 (LF.IX), ML353 (HF.IX), ML354-355 (LF.IX), ML356 (HF.IX), ML357-363 (LF.IX), ML364 (HF.IX), ML365-366 (LF.IX), ML367 (HF.IX), ML368-381 (LF.IX), ML396-398 (LF.IX), ML399 (HF.IX), ML400-404 (LF.IX), ML405-406 ((HF.IX), ML407-412 (LF.IX), ML413 (HF.IX), ML414-424 (LF.IX), ML425 (HF.IX) and ML426-428 (LF.IX).

Converted to HF.IX: MK670, MK672, MK674-675, MK680-683, MK691, MK694, MK696-697, MK842, MK965, ML145, ML148, ML150, ML154, ML156, ML171, ML174, ML179, ML182, ML185-187, ML195-197, ML207, ML209-212 and ML216.

Converted to LF.XVI: MJ556 (prototype).

Spitfire LF.IX MJ238.

MT502-MV514

700 Spitfire VIIIs ordered in July 1942 and built as 585 MK.VIII and 115 Mk.XIV between December 1943 and October 1944. Most delivered direct to India and Australia.

Serial numbers: MT502-507 (LF.VIII), MT539-581 (LF.VIII), MT595-635 (LF.VIII), MT648-674 (LF.VIII), MT675 (HF.VIII), MT676-683 (LF.VIII), MT684 (HF.VIII), MT685-689 (LF.VIII), MT703-747 (LF.VIII), MT748 (HF.VIII), MT761-802 (LF.VIII), MT815 (LF.VIII), MT816-817 (HF.VIII), MT818 (LF.VIII), MT819-822 (HF.VIII), MT823-824 (LF.VIII), MT825 (HF.VIII), MT826-828 (LF.VIII), MT829-831 (HF.VIII), MT832 (LF.VIII), MT833-835 (HF.VIII), MT836-846 (LF.VIII), MT847-858 (FR.XIV), MT872-889 (LF.VIII), MT890-900 (HF.VIII), MT901-909 (LF.VIII), MT910 (HF.VIII), MT911-913 (LF.VIII), MT914 (HF.VIII), MT915 (LF.VIII), MT925-969 (LF.VIII), MT981-999 (LF.VIII).

MV112-117 (HF.VIII), MV118 (LF.VIII), MV119-121 (HF.VIII), MV122 (LF.VIII), MV123-125 (HF.VIII), MV126-127 (LF.VIII), MV128-129 (HF.VIII), MV130-131 (LF.VIII), MV132-133 (HF.VIII), MV134-140 (LF.VIII), MV141 (HF.VIII), MV142-143 (LF.VIII), MV144-156 (HF.VIII), MV169-208 (HF.VIII), MV231-232 (LF.VIII), MV233 (HF.VIII), MV234 (LF.VIII), MV235 (HF.VIII),

Spitfire LF.IX MK264.

MV236 (LF.VIII), MV237 (HF.VIII), MV238 (LF.VIII), MV239-244 (HF.VIII), MV245 (LF.VIII), MV246-273 (F.XIV), MV286-320 (F.XIV), MV321-325 (HF.VIII), MV326-329 (LF.VIII), MV342 (HF.VIII), MV343 (LF.VIII), MV344 (HF.VIII), MV345 (LF.VIII), MV346 (HF.VIII), MV347-386 (F.XIV), MV398-441 (LF.VIII), MV456-459 (LF.VIII), MV460-461(HF.VIII), MV462 (LF.VIII), MV463-479 (HF.VIII), MV480 (LF.VIII), MV481-482 (HF.VIII), MV483 (LF.VIII), MV484-487 (HF.VIII) and MV500-514 (HF.VIII).

NG757-NH611

600 Spitfire IXs ordered from Castle Bromwich in December 1942 and built as 324 LF.IX and 44 HF.IX between April and June 1944, remainder (in the NG757-NH147 range) cancelled.

Serial numbers: NH148-158, NH171-218, NH230-276, NH289-326, NH339-381, NH393-438, NH450-496, NH513-558 and NH570-611.

The HF.IXs were: NH148, NH153, NH181, NH190, NH194, NH236, NH250, NH256, NH262, NH267, NH271, NH275, NH293, NH297, NH310, NH313, NH360, NH362, NH418, NH420, NH422, NH433, NH437, NH450, NH459, NH478, NH482, NH486, NH488, NH513, NH518, NH528, NH534, NH536, NH542, NH545, NH547, NH572, NH577, NH578, NH582, NH587 and NH611.

NH614-NH929
225 Spitfire VIIIs ordered December 1942 and built as 23 Mk.VIII, 191 Mk.XIV and 11 Mk.XVIII between November 1944 and June 1945.
Serial numbers: NH614 (HF.VIII), NH615-636 (LF.VIII), NH637-661 (FR.XIV), NH685-720 (FR.XIV), NH741-759 (FR.XIV), NH775-789 (FR.XIV), NH790 (F.XIV), NH791-796 (FR.XIV), NH797 (F.XIV), NH798-811 (FR.XIV), NH812 (F.XIV), NH813 (FR.XIV); NH831-846 (F.XIV), NH847-856 (FR.XVIII), NH857-871 (F.XIV), NH872 (FR.XVIII), NH873-875 (F.XIV) and NH892-929 (F.XIV).

NM814-NM906
144 Spitfire VIIIs ordered December 1942 in serial ranges NM775-832 and NM879-906 but cancelled. New order for 10 Spitfire FR.XIVs placed with serials NM814-823 and built March-April 1945.

PA838-PA961
100 Spitfire VIIIs ordered April 1943 and built as 93 PR.XI and 7 LF.VIII between November 1943 and March 1944.
Serial numbers: PA838-871 (PR.XI), PA884-913 (PR.XI), PA926-951 (PR.XI), PA952-958 (LF.VIII), PA959-961 (PR.XI).

Spitfire F.22 PK312.

PK312-PL499
Originally an order for 800 Spitfire LF.IXs placed with Castle Bromwich in June 1943, this contract was changed several times and resulted in 571 aircraft being built between June 1944 and December 1945 comprising 284 L/LF.IX, 260 F.22 and 27 F.24; remainder cancelled.
Serial numbers: PK312-356 (F.22), PK369-412 (F.22), PK426-435 (F.22), PK481-525 (F.22), PK539-582 (F.22), PK594-635 (F.22), PK648-677 (F.22), PK678-689 (F.24), PK712-726 (F.24), PK991-998 (LF.IX).
PL123-169 (LF.IX), PL185-188 (LF.IX), PL189 (HF.IX), PL190-191 (LF.IX), PL192 (HF.IX), PL193 (LF.IX), PL194 (HF.IX), PL195-202 (LF.IX), PL203 (HF.IX), PL204-208 (LF.IX), PL209 (HF.IX), PL210-214 (LF.IX), PL215 (HF.IX), PL216-217 (LF.IX), PL218 (HF.IX), PL219-221 (LF.IX), PL222 (HF.IX), PL223-227 (LF.IX), PL246-248 (LF.IX), PL249 (HF.IX), PL250-252 (LF.IX), PL253 (HF.IX), PL254-255 (LF.IX), PL256 (HF.IX), PL257-260 (LF.IX), PL261 (HF.IX), PL262 (LF.IX), PL263 (HF.IX), PL264-288 (LF.IX), PL313-356 (LF.IX), PL369-370 (LF.IX), PL371-372 (HF.IX), PL373 (LF.IX), PL374-380 (HF.IX), PL381-383 (LF.IX), PL384-388 (HF.IX), PL389 (LF.IX), PL390 (HF.IX), PL391 (LF.IX), PL392 (HF.IX), PL393-395 (LF.IX), PL396 (HF.IX), PL397-399 (LF.IX), PL400 (HF.IX), PL401-408 (LF.IX), PL423-431 (LF.IX), PL432 (HF.IX), PL433-449 (LF.IX), PL450 (HF.IX), PL451 (LF.IX), PL452 (HF.IX), PL453-466 (LF.IX), PL488 (LF.IX), PL489 (HF.IX) and PL490-499 (LF.IX).

PL758-PM676
Many production contracts for Spitfires were cancelled or changed more than once towards the end of the war,

those covered by this serial range one of them. It started as a June 1942 order for 592 Spitfire VIIIs, was modified to include some Spitefuls and after several other amendments finally resulted in 227 PR.XIs and 124 PR.XIXs being built between February 1944 and May 1946, a total of 351 aircraft.
Serial numbers: PL758-799 (PR.XI), PL823-866 (PR.XI), PL881-925 (PR.XI), PL949-998 (PR.XI), PM123-168 (PR.XI), PM462 (PR.XIX), PM496-519 (PR.XIX), PM536-581 (PR.XIX), PM596-637 (PR.XIX) and PM651-661 (PR.XIX).

PP139
The serial PP139 was allocated to the production prototype Spitfire F.21 first flown July 1943. Due to an administrative mixup, the same serial number adorned a Short Sunderland III flying boat.

PS684-PS935
200 Spitfire VIIIs ordered July 1943, later amended to cover 121 Spitefuls and 79 Spitfire PR.XIXs but finally built as the 79 PR.XIXs only between November 1944 and May 1945.
Serial numbers: PS831-836, PS849-893 and PS908-935.

PS936-PW250
1,500 Spitfire 21s ordered from Castle Bromwich July 1943. Contract cancelled and amended to cover 673 Spitfire IXs and subsequently 200 Griffon powered Seafires, these also cancelled. Final makeup was 594 Spitfire LF.IXs and 79 HF.IXs (total 673, serials below) built June-October 1944 and 14 Seafire 47s (PS944-957, see Seafire section).
Serial numbers: PT355-380, PT395-436, PT451-498, PT523-567, PT582-627, PT639-683, PT697-738, PT752-795, PT818-859, PT873-915, PT929-970, PT986-999, PV115-160, PV174-215, PV299-270, PV283-327 and PV341-359, remainder cancelled.
Completed as HF.IX: PT398, PT432, PT434, PT455, PT460, PT462-463, PT465-466, PT470, PT473-474, PT480-481, PT486, PT488, PT493, PT601, PT605, PT608, PT612, PT614, PT619, PT627, PT640, PT650, PT657, PT714, PT733, PT753, PT756, PT760-761, PT764-766, PT768, PT781, PT787, PT818, PT835, PT847, PT876, PT888, PT903-905, PT907, PT910, PT913, PT915, PT929, PT931-932, PT941, PV229, PV232, PV238, PV259, PV261, PV264, PV269, PV283-284, PV286, PV290, PV296, PV299, PV303-304, PV308, PV312, PV318, PV321, PV324, PV343-344 and PV346.
Converted to LF.XVI: PV288, PV295, PV307, PV327 and PV349.

RB140-RB189
50 Spitfire F.XIVs ordered August 1943 and built between October 1943 and March 1944. Serial RB140 clashed with the same number allocated to a Slingsby Cadet glider.
Serial numbers: RB140-189.

Spitfire F.XIV RB159.

RB515-RB987
373 Spitfire F.21s ordered August 1943 but all cancelled. 19 built as Spitefuls (RB515-525, RB527-531, RB533-535).

RK798-RK926
100 Spitfire IXs ordered from Castle Bromwich in October 1943 as built as 65 LF.IX, 10 HF.IX and 25 LF.XVI between August and October 1944. Some to Soviet Union.

Serial numbers: RK798-810 (LF.IX), RK811 (HF.IX), RK812-819 (LF.IX), RK835-839 (LF.IX), RK840 (LF.XVI), RK841 (LF.IX), RK842 (LF.XVI), RK843-848 (LF.IX), RK849 (LF.XVI), RK850-858 (LF.IX), RK859 (LF.XVI), RK860 (HF.IX), RK861-864 (LF.IX), RK865-866 (LF.XVI), RK867 (LF.IX), RK868 (LF.XVI), RK883 (LF.XVI), RK884-887 (LF.IX), RK888 (LF.XVI), RK889 (HF.IX), RK890 (LF.IX), RL891-893 (LF.XVI), RK894 (LF.IX), RK895-897 (LF.XVI), RK898-900 (LF.IX), RK901 (HF.IX), RK902-905 (LF.XVI), RK906-907 (LF.IX), RK908 (HF.IX), RK909 (LF.IX), RK910 (LF.XVI), RK911-912 (HF.IX), RK913 (LF.XVI), RK914-915 (LF.IX), RK916-917 (HF.IX), RK918 (LF.XVI), RK919-920 (LF.IX), RK921 (LF.XVI), RK922-923 (LF.IX), RK924 (HF.IX), RK925-926 (LF.XVI).

Spitfire F.XIVE RM784.

RM615-RN221
406 Spitfire XIVs ordered October 1943 and built as 384 F/FR.XIV/XIVE and 22 PR.XIX between April 1944 and May 1945.

Serial numbers: RM615-625 (F.XIV), RM626-647 (PR.XIX), RM648-656 (F.XIV), RM670-713 (F.XIV), RM726-770 (F.XIV), RM783-785 (F.XIV), RM786 (FR.XIV), RM787-825 (F.XIV), RM839-887 (F.XIV), RM901-943 (F.XIV), RM957-999 (F.XIV), RN113-160 (F.XIV), RN173-203 (F.XIV), RN204 (FR.XIV), RN205-216 (F.XIV), RN217-221 (FR.XIV).

RR181-RR265
73 Spitfire IXs ordered November 1943 and 72 built as 31 LF.IX, 19 HF.IX and 22 LF.XVI between August and October 1944.

Serial numbers: RR181-205 (LF.IX), RR206 (HF.IX), RR207-208 (LF.IX), RR209 (HF.IX), RR210-211 (LF.IX), RR212-213 (LF.XVI), RR226-227 (LF.XVI), RR228 (HF.IX),

Spitfire F.XIVE RM825.

Spitfire HF.IX RR239.

RR229-230 (LF.XVI), RR231-232 (HF.IX), RR233 (cancelled), RR234 (LF.XVI), RR235 (HF.IX), RR236 (LF.XVI), RR237 (LF.IX), RR238-239 (HF.IX), RR240 (LF.XVI), RR241 (HF.IX), RR242-243 (LF.XVI), RR244 (HF.IX), RR245 (LF.XVI), RR246 (HF.IX), RR247-250 (LF.XVI), RR251-252 (HF.IX), RR253 (LF.IX), RR254 (HF.IX), RR255-257 (LF.XVI), RR258-260 (HF.IX), RR261 (LF.XVI), RR262 (HF.IX), RR263 (LF.XVI), RR264 (HF.IX) and RR265 (LF.XVI).

RV370-RX151
700 Spitfire 21s ordered from Castle Bromwich in January 1944 but order much and amended large sections cancelled. Finally completed as only 40 Spitfire LF.XVI built June-July 1945.

Serial numbers: RW344-359 and RW373-396, remainder cancelled.

SL541-SM698
800 Spitfire 21s ordered from Castle Bromwich in February 1944 but this order also much amended. Finally completed as 179 LF.IX and 379 LF.XVI (total 558) built between September and November 1944. 24 LF.IXs to Czech Air Force and approximately 105 LF.IXs to Soviet Union.

Spitfire LF.XVIE SM256.

Serial numbers: SL541-579 (LF.XVI), SL593-624 (LF.XVI), SL625-635 (LF.IX), SL648-665 (LF.IX), SL666-690 (LF.XVI); SL713-745 (LF.XVI), SM135-150 (LF.IX), SM170-177 (LF.IX), SM178-213 (LF.XVI), SM226-258 (LF.XVI), SM273-316 (LF.XVI), SM329-369 (LF.XVI), SM383-427 (LF.XVI), SM441-463 (LF.IX), SM464-488 (LF.XVI), SM503-516 (LF.XVI), SM517-548 (LF.IX), SM563-566 (LF.IX), SM567 (LF.XVI), SM568-597 (LF.IX), SM610-645 (LF.IX), SM646-648 (LF.XVI) and SM663-671 (LF.XVI), remainder cancelled.

SM812-SM997
150 Spitfire PR.XIs ordered February 1944 and built as 31 F.XIV, 50 FR.XIV and 18 F.XVIII and 33 FR.XVIII (total 132) between February 1945 and January 1946. 2 F.XVIIIs to Indian Air Force.

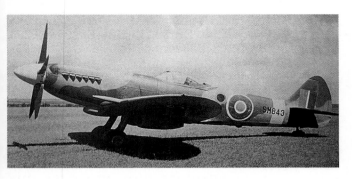

Spitfire FR.18 SM843.

Serial numbers: SM812-842 (F.XIV), SM843-845 (FR.XVIII), SM876-899 (FR.XIV), SM913-938 (FR.XIV), SM939-956 (F.XVIII) and SM968-997 (F.XVIII).

SR395-SR400

6 Spitfire VIIs ordered February 1944 and built as PR.Xs, delivered from May 1944.
Serial numbers: SR395-400.

SW777

1 prototype Spitfire 21 ordered March 1944 but built as prototype pressurised version of Spitfire PR.XIX.

SX549

Prototype Spitfire 21 ordered April 1944 but built as the prototype F.22 in November 1944.

TA738-TE579

1,884 Spitfire IXs ordered from Castle Bromwich in April 1944 of which 1,482 built between December 1944 and June 1945 as 740 LF.IX, 100 HF.IX and 642 LF.XVI. Approximately 275 LF.IXs to Soviet Union.
Serial numbers: TA738-780 (LF.IX), TA793 (LF.IX), TA794-796 (HF.IX), TA797 (LF.IX), TA798 (HF.IX), TA799 (LF.IX), TA800 (HF.IX), TA801 (LF.IX), TA802 (HF.IX), TA803 (LF.IX), TA804-808 (HF.IX), TA809 (LF.IX), TA810-813 (HF.IX), TA814-816 (LF.IX), TA817 (HF.IX), TA818-824 (LF.IX), TA825 (HF.IX), TA826-840 (LF.IX), TA850-888 (LF.IX), TA905-948 (LF.IX), TA960-999 (LF.IX). TB115-129 (LF.IX), TB130-132 (LF.XVI), TB133-135 (LF.IX), TB136-141 (LF.XVI), TB142-150 (LF.IX), TB168-193 (LF.IX), TB194 (cancelled), TB195-197 (LF.IX), TB213-243 (LF.IX), TB244-248 (LF.XVI), TB249-251 (LF.IX), TB252 (LF.XVI), TB253 (LF.IX), TB254-256 (LF.XVI), TB269-308 (LF.XVI), TB326-349 (LF.XVI), TB352-396 (LF.XVI), TB413-450 (LF.IX), TB464-474 (LF.IX), TB475-476 (LF.XVI), TB477 (LF.IX), TB478 (LF.XVI), TB479 (LF.IX), TB480-481 (LF.XVI), TB482-491 (LF.IX), TB492-502 (LF.XVI), TB503 (LF.IX), TB515 (LF.XVI), TB516-518 (LF.IX), TB519-522 (LF.XVI), TB523-524 (LF.IX), TB525-526 (LF.XVI), TB527 (LF.IX), TB528 (LF.XVI), TB529-531 (LF.IX), TB532 (HF.IX), TB533-536 (LF.IX), TB537 (HF.IX), TB538-541 (LF.IX), TB542 (HF.IX), TB543 (LF.IX), TB544-548 (HF.IX), TB563 (LF.IX), TB564-565 (HF.IX), TB566-567 (LF.IX), TB568-570 (HF.IX), TB571 (LF.IX), TB572-574 (LF.XVI), TB575 (LF.IX), TB576-577 (HF.IX), TB578 (LF.XVI), TB579 (LF.IX), TB580-583 (LF.XVI), TB584 (HF.IX), TB585 (LF.XVI), TB586-587 (HF.IX), TB588-590 (LF.XVI), TB591-598 (LF.IX), TB613-637 (LF.XVI), TB638-659 (LF.IX), TB674 (LF.IX), TB675 (LF.XVI), TB676-712 (LF.IX), TB713-716 (LF.XVI), TB717-718 (LF.IX), TB733-735 (LF.XVI), TB736 (LF.IX), TB737-739 (LF.XVI), TB740 (LF.IX), TB741-759 (LF.XVI), TB771-809 (LF.IX),

TB824-827 (LF.IX), TB828-829 (LF.XVI), TB830 (LF.IX), TB831-836 (LF.XVI), TB837-843 (LF.IX), TB844-847 (HF.IX), TB848-857 (LF.IX), TB858-868 (LF.XVI), TB883-900 (LF.XVI), TB901-917 (LF.IX), TB918 (HF.IX), TB919 (LF.XVI), TB920 (HF.IX), TB921-923 (LF.XVI), TB924-925 (HF.IX), TB938-959 (LF.IX), TB971-980 (LF.IX), TB981-988 (HF.IX), TB989-991 (LF.XVI), TB992 (HF.IX), TB993 (LF.XVI), TB994 (HF.IX), TB995-999 (LF.XVI).
TD113-154 (LF.XVI), TD155 (LF.IX), TD156-158 (LF.XVI), TD175 (LF.IX), TD176-177 (LF.XVI), TD178-183 (LF.IX), TD184-191 (LF.XVI), TD192-204 (LF.IX), TD205-206 (HF.IX), TD207-213 (LF.IX), TD229-267 (LF.XVI), TD280-286 (LF.XVI), TD287 (LF.IX), TD288-289 (LF.XVI), TD290-292 (LF.IX), TD293 (LF.XVI), TD294-304 (LF.IX), TD305 (HF.IX), TD306-309 (LF.IX), TD310 (HF.IX), TD311-312 (LF.IX), TD313-315 (HF.IX), TD316-325 (LF.XVI), TD338-351 (LF.XVI), TD352 (LF.IX), TD353-354 (HF.IX), TD355 (LF.IX), TD356-359 (HF.IX), TD360 (LF.IX), TD361-363 (HF.IX), TD364-365 (LF.IX), TD366-367 (HF.IX), TD368 (LF.IX), TD369 (LF.XVI), TD370 (HF.IX), TD371 (LF.IX), TD372 (LF.XVI), TD373-374 (LF.IX), TD375-377 (LF.XVI), TD378-379 (LF.IX), TD395-399 (LF.IX), TD400-408 (LF.XVI), TD952-958 (LF.IX), TD970-980 (LF.IX), TD981-999 (LF.IX).
TE115 (LF.IX), TE116 (LF.XVI), TE117-118 (LF.IX), TE119-120 (LF.XVI), TE121-158 (LF.IX), TE174-196 (LF.XVI), TE197 (HF.IX), TE198-204 (LF.XVI), TE205 (HF.IX), TE206-210 (LF.XVI), TE211-213 (HF.IX), TE214 (LF.XVI), TE215 (HF.IX), TE228-229 (LF.XVI), TE230-234 (HF.IX), TE235-237 (LF.XVI), TE238 (HF.IX), TE239-259 (LF.XVI), TE273-291 (LF.XVI), TE292-299 (HF.IX), TE300 (LF.XVI), TE301-309 (HF.IX), TE310-311 (LF.XVI), TE312-313 (HF.IX), TE314 (LF.XVI), TE315 (HF.IX), TE328 (LF.XVI), TE329 (HF.IX), TE330 (LF.IX), TE331 (HF.IX), TE332 (LF.XVI), TE333 (HF.IX), TE334-335 (LF.XVI), TE336-337 (HF.IX), TE338-342 (LF.XVI), TE343 (LF.IX), TE344-359 (LF.XVI), TE375-385 (LF.XVI), TE386 (cancelled), TE387-408 (LF.XVI), TE434-471 (LF.XVI), TE472 (cancelled), TE473-480 (LF.XVI), TE493-535 (LF.IX) and TE549-579 (LF.IX).

TL773-TM251

276 Spitfire IXs ordered from Castle Bromwich in July 1944; all but one (TM115 as a Spitfire XVI) cancelled October 1944, TM115 also cancelled August 1945.

TP195-TP456

300 Spitfire XIVs ordered August 1944 but built as 200 FR.XVIII and 6 F/FR.XIV (total 206) between December 1945 and February 1946.
Serial numbers: TP195-235 (FR.XVIII), TP236-240 (F.XIV), TP256 (FR.XIV), TP257-298 (FR.XVIII), TP313-350 (FR.XVIII), TP363-408 (FR.XVIII) and TP432-456 (FR.XVIII), remainder cancelled.

TX974-TZ240

157 Spitfire XIVs ordered February 1945 and built as 25 F.XIV, 95 FR.XIV and 37 FR.XVIII between August 1945 and January 1946.
Serial numbers: TX974-998 (F.XIV), TZ102-149 (FR.XIV), TZ152-176 (FR.XIV), TZ178-188 (FR.XIV), TZ200-205 (FR.XVIII) and TZ210-240 (FR.XVIII).

TZ598-TZ738

100 Spitfire PR.XIXs ordered March 1945 but all cancelled mid 1945.

TZ747-VA195

188 Spitfire LF.XVIs ordered from Castle Bromwich in March 1945 but all cancelled August 1945.

VA201-VA250

50 Spitfire F.22s ordered from Castle Bromwich in March 1945 but cancelled August 1945.

VN301-VN496

150 Spitfire F.22s ordered from Castle Bromwich in November 1945 as the final Spitfire contract and included using up 54 incomplete airframes. These were the only ones built (as Spitfire F.24s) between April 1946 and February 1948, remainder cancelled. Production transferred from Castle Bromwich to the Vickers-Supermarine facility at South Marston.
Serial numbers: VN301-334 and VN477-496.
Note: VN496 was the last Spitfire delivered to the RAF, on 20 February 1948.

SEAFIRE

LA428-LA564

Cancelled Spitfire 21 order reinstated to cover 50 Seafire F.45s (built at Castle Bromwich, delivered March 1945-January 1946) and 24 Seafire F.46s (built at South Marston, delivered November 1945-May 1947).
Serial numbers: LA428-462 (F.45), LA481-495 (F.45) and LA541-564 (F.46).

Seafire 45 LA429.

LR631-LR881

213 Seafire IICs ordered from Westland Aircraft in March 1942 and built as 110 Mk.IIC and 103 MK.III between January 1943 and January 1944.
Serial numbers: LR631-667 (IIC), LR680-712 (IIC), LR725-764 (IIC), LR765-820 (III) and LR835-881 (III).

MA970-MB327

202 Seafire IICs ordered March 1942 as conversions of Spitfire VCs on production line, built between June 1942 and April 1943.
Serial numbers: MA970-999, MB113-158, MB178-222, MB235-281 and MB293-326.
Conversion: MA970 to prototype Seafire III (first flight after conversion 9 November 1942).

MB328-MB375

48 Seafire IBs ordered from Air Service Training as conversions of Spitfire VBs and given new serial numbers between MB328 and MB375. Delivered mid 1942.
Previous identities were (in order MB328-375): BL676, BL687, BL678, BL694, AB416, AB410, AB413, AB408, AB376, AB261, AB415, BL679, BL689, AB414, AB379, AB409, AR344, AR445, AR446, AR443, AR459, AR442, AB404, AB405, AB406, AB407, AB492, AB408, EP148,

AR457, AR458, EP141, AR460, AR461, EP142, EP144, EP146, EP147, EP291, EP293, EP294, EP295, EP296, EP299, EP301, EP302, EP304 and EP308.

MV660-MV990

260 Seafires ordered July 1942 but cancelled August 1942.

NF418-NF665

200 Seafire IICs ordered from Westland Aircraft in November 1942 built as 38 LF.IIC and 162 L.III, first delivery January 1944.
Serial numbers: NF418-455 (IIC), NF480-526 (III), NF531-570 (III), NF576-607 (III) and NF624-665 (III).
Note: 26 aircraft in NF418-455 block were built as Seafire IIIs but with non folding wings, these originally designated LF.III (Hybrid) and then LF.IIC (Hybrid).

NM910-NM982

60 Seafire IICs ordered January 1943, converted from Spitfire VCs EN686-695 and EN710-759 on production line and allocated new serial numbers. Built January-March 1943.
Serial numbers: NM910-949 and NM963-982.

NM984-NN330

200 Seafire IIIs ordered from Westland Aircraft in January 1943 and built between April and August 1944.
Serial numbers: NM984-999, NN112-157, NN169-214, NN227-270 and NN283-330.
Note: small number built as Seafire LF.IIC (Hybrid) with non folding wings.

NN333-NN641

250 Seafire IIIs ordered from Cunliffe-Owen Aircraft in January 1943 and built between December 1943 and November 1944.
Serial numbers: NN333-367, NN379-418, NN431-476, NN488-528, NN542-586 and NN599-641.

NS487, NS490 and NS493

3 Seafire XV prototypes ordered March 1943 and built late 1943. NS493 modified to Seafire XVII standards.

NX879-PA129

118 Seafire IB conversions (from Spitfire VBs) ordered from Cunliffe-Owen Aircraft in late 1942 and delivered February-June 1943.
Serial numbers: NX879-928, NX940-967, NX980-999 and PA100-129.

PK240, PK243 and PK245

3 Seafire XV prototypes ordered in May 1943 and built early 1944.

PP921-PR334

250 Seafire IIIs ordered from Westland Aircraft in July 1943 and built between August and December 1944.
Serial numbers: PP921-957, PP969-999, PR115-156, PR170-215, PR228-271 and PR285-334.

PR338-PR506

150 Seafire IIIs ordered from Cunliffe-Owen Aircraft in July 1943 and built as 134 Seafire XVs between December 1944 and May 1945.
Serial numbers: PR338-379, PR391-436, PR449-479 and PR492-506.
Note: 11 delivered to Burmese Airforce and 7 to Royal Canadian Navy.

Seafire FR.47 PS944

PS944-PS957

Part of the much amended contract for 1,500 Spitfire 21s placed with Castle Bromwich in July 1943 (PS936-PW250), this section (PS944-957) covering 14 Seafire 47s ordered in 1945 and built at South Marston between June 1946 and November 1947.

PV734-PW122

200 Griffon-Seafires ordered from Castle Bromwich under same circumstances as PS944-957 (above) but cancelled August 1945.

RX156-RX530

300 Seafire IIIs ordered from Westland Aircraft in January 1944 of which 160 built between December 1944 and July 1945, remainder cancelled March 1944.
Serial numbers: RX156-194, RX210-256, RX268-313 and RX326-353.

SP136-SP355

200 Seafire XVs ordered from Cunliffe-Owen Aircraft in February 1944 but reduced to cover 50 Seafire XV and 20 Seafire XVII built June-December 1945.
Serial numbers: SP136-168 (XV), SP181-197 (XV), SP323-327 (XVII) and SP341-355 (XVII).

SR446-SR645

140 Seafire XVs ordered from Westland Aircraft in February 1944 and built between August 1944 and April 1945.
Serial numbers: SR446-493, SR516-547, SR568-611 and SR630-645.
Note: 6 delivered to Burmese Air Force.

Seafire F.XVIIs SX194 (above) and SX358 (below).

SW781-SX546

503 Seafire XVs ordered from Westland Aircraft in March 1944 and built as 110 F.XV and 212 F.XVII between April 1945 and October 1946, remainder cancelled.
Serial numbers: SW781-828 (XV), SW844-875 (XV), SW876-879 (XV), SW896-921 (XV), SW986-993 (XVII), SX111-139 (XVII), SX152-201 (XVII), SX220-256 (XVII), SX271-316 (XVII), SX332-370 (XVII) and SX386-389 (XVII).
Note: 3 F.XVs delivered to Burmese Air Force.

TM379 and TM383

3 Spitfire 21s ordered from Castle Bromwich in May 1944 but amended to cover two Seafires as F.45 (TM379) and F.46 (TM383) prototypes. Third aircraft (TM389) cancelled.

VD490-VE593

600 Seafire 45/46/47s ordered April 1945 but all cancelled August 1945.

VN501-VN673

135 Seafire 46/47s ordered November 1945 but all cancelled December 1945.

Seafire FR.47 VP482.

VP427-VP495

64 Seafire F.47s ordered in April 1946 and built between November 1947 and November 1948.
Serial numbers: VP427-465 and VP471-495.

VR961-VR972

12 Seafire F.47s ordered in October 1946 and delivered between November 1948 and January 1949. VR972 was the last of the Spitfire/Seafire line, delivered on 28 January 1949.

RAF SPITFIRE SQUADRONS

RAF SPITFIRE SQUADRONS

Note: The table summarises the Royal Air Force operational squadrons which operated Spitfires between 1938 and 1954 when regular service ended. It includes Commonwealth units operating as part of the RAF plus those squadrons manned to a greater or lesser extent by personnel (mainly pilots) from the European countries occupied by Germany throughout WWII.

The table's headings list the squadron number, the years it operated the Spitfire, code letters/numbers, the marks of Spitfire operated and areas of operations. For convenience, Arabic numerals are used for the various Spitfire marks, even though this method of appellation was not adopted by the RAF until after the war. Before that, Roman numerals were used.

Abbreviations: Aust – Australia; ETO – European Theatre of Operations; FE – Far East (Burma, Cocos Islands, Hong Kong, India, Japan, Malaya, New Guinea, Pacific etc); ME – Middle East (Mediterranean, Italy, North Africa, Palestine, Southern France, Gibraltar etc); RAAF – Royal Australian Air Force; RCAF – Royal Canadian Air Force; RNZAF – Royal New Zealand Air Force; UK – United Kingdom.

Sqdn	Mark/s	Dates	Code/s	Notes/Operations
1	9/21	1944-46	JX	UK
2	11/14/19	1945-51	OI	ETO, tactical reconnaissance
4	11/13/19	1943-45	-	ETO, tactical reconnaissance
5	16	1949-51	7B	UK, anti aircraft co-operation
6	9	1945-46	JV	ME
11	14/18	1945-48	-	FE
16	11/14/16/19	1943-46	-	ETO, tactical reconnaissance
17	8/14/16	1944-51	YB/5S/UT	FE/UK
19	1/2/5/9	1938-44	19/WZ/QV	UK
20	8/14	1945-46	HN	FE
26	5/11/14	1944-47	XC	ETO, tactical reconnaissance
28	8/9/14/18	1945-51	-	FE
32	5/8/9/18	1943-49	GZ	ME
33	5/9/16	1943-46	NW/5R	ME/UK
34	16	1949-51	4M/8Q	UK, anti aircraft co-operation
41	1/2/5/12/14/21	1939-47	PN/EB	UK/ETO/UK
43	5/8/9	1943-47	FT	ME
54	1/2/5/8	1939-45	DL/KL	UK/Aust
56	9	1944	US	UK
60	18/19	1947-50	-	FE
63	5/16	1944-48	UB	UK
64	1/2/5/9	1940-44	GR	UK
65	1/2/5/9/16	1939-46	FZ/YT	UK
66	1/2/5/9/16	1938-47	RB/LZ	UK/ETO/UK
67	8	1944-45	RD	FE
69	4	1941-43	-	ME, photo-reconnaissance
71	2/5	1941-42	XR	UK, Eagle Squadron became 334th FS
72	1/2/5/9	1939-46	RN	UK/ME
73	5/8/9/22	1943-48	-	ME
74	1/2/5/9/16	1939-45	JH/ZP/4D	UK/ME/UK
80	5/9/24	1943-52	EY/W2	UK/ETO/FE
81	5/8/9/18/19	1942-54	FL	UK/ME/FE, fighter and photo-recce
87	5/8/9	1943-46	LK	ME
91	2/5/9/12/14/21	1941-46	DL	UK
92	1/5/8/9	1940-46	GR/QJ	UK/ME
93	5/9	1942-45	HN	UK/ME
94	5/8/9	1944-45	GO	ME
111	1/2/5/9	1941-47	JU	UK/ME
118	2/5/9	1941-45	NK	UK
121	2/5	1941-42	AV	UK, Eagle Squadron became 335th FS
122	1/2/5/9*	1941-44*	MT	UK (* also Mk.21 in 1946)
123	1/2/5	1941-43	XE	UK/ME
124	1/2/5/6/7/9	1941-45	ON	ETO
126	5/9*	1942-44*	TD	ME/UK (* also Mk.16 in 1946)
127	5/9/16	1943-45	9N	ME/UK
129	1/2/5/6/9	1941-46	DV	UK
130	2/5/9/14	1941-46	PJ/AP	UK/ETO
131	1/2/5/7/8/9	1941-45	NX	UK/FE
132	1/2/5/8/9/14	1941-46	FF	ETO/FE
133	2/5/9	1941-42	MD	UK, Eagle Squadron became 336th FS
134	2/5	1941-43	GQ	UK/ME
136	5/8/14	1943-46	HM	FE
140	4/11	1941-44	-	UK, photo-reconnaissance
145	2/5/8/9	1941-45	SO/ZX	UK/ME
152	1/2/5/8/9/14	1940-46	SN/UM	UK/ME/FE
154	2/5/8/9	1941-45	UO/HT	UK/ME
155	8	1944-46	DG	FE
164	5/9/16	1942-46*	FJ	UK/ETO (* flew Hurricanes/Typhoons 1943-45)
165	5/9	1942-46	SK	UK/ETO
167	5	1942-43	VL	UK
185	5/8/9	1942-45	GL	ME
186	5	1944	-	UK
208	5/8/9/18	1943-51	RG	ME
212	1 (PR)	1940	-	UK, photo-reconnaissance
213	5/9	1944	AK	ME
222	1/2/5/9	1940-44	ZD	ETO
225	5/9	1943-47	WU	ME
229	5/9/16	1942-45	9R	ME/ETO
232	5/9	1942-44	EF	UK/ME
234	1/2/5/6/9*	1940-46	AZ/FX	UK (* flew Mustangs 1944-45)
237	5/9	1943-45	DV	ME
238	1/5/8/9	1940-44*	KC	UK/ME (* flew Hurricanes 1940-42)
241	5/8/9	1943-45	RZ	UK/ME
242	5/9	1942-44	LE	UK/ME
243	5/9	1942-44	SN	UK/ME
249	5/9*	1942-45*	GN	ME (* also Mk.1 1940 and Mustang 1944-45)
253	5/8/9/11	1943-47	SW	ME
266	1/2/5	1940-42	ZH	UK
268	14/19	1945	-	ETO, tactical reconnaissance
269	5	1944-46	UA	UK, air-sea rescue
273	8/14	1944-46	MS	FE
274	5/9	1943-44	NH/JJ	ME/UK
275	5	1943-45	PV	UK, air-sea rescue
276	2/5	1943-45	VA	UK/ETO, air-sea rescue
277	2/5	1942-45	BA	UK, air-sea rescue
278	2/5	1942-45	MY	UK, air-sea rescue
287	5/9/16	1943-46	KZ	UK, anti aircraft co-operation
288	9/16	1944-46*	RP	UK, anti aircraft co-operation (*also Mk.16 1953)
289	5	1944-45	YE	UK, anti aircraft co-operation

Sqdn	Mark/s	Dates	Code/s	Notes/Operations
290	5	1944-45	X6	UK, anti aircraft co-operation
302	5/9/16	1941-46	WX	UK/ETO (Polish)
303	1/2/5/9/16	1941-45	RF	UK (Polish)
306	2/5/9	1941-43	UZ	UK (Polish)
308	1/2/5/9/16	1941-46	ZF	UK/ETO (Polish)
310	2/5/6/9	1941-46	NN	UK/ETO (Czech)
312	2/5/9	1941-46	DU	UK/ETO (Czech)
313	1/2/5/6/7/9	1941-46	RY	UK/ETO (Czech)
315	2/5/9	1941-43	PK	UK (Polish)
316	5/9	1941-43	SZ	UK (Polish)
317	5/9/16	1941-46	JH	UK/ETO (Polish)
318	5/9	1944-46	LW	ME (Dutch)
322	5/9/14/16	1943-45	3W	UK/ETO (Dutch)
326	5/9	1943-45	9I	ETO (Free French)
327	5/8/9	1943-45	-	ETO (Free French)
328	5/8/9	1943-45	S8	ME/ETO (Free French)
329	5/9/16	1944-45	5A	UK/ETO (Free French)
331	2/5/9	1941-45	FN	UK/ETO (Norwegian)
332	5/9	1942-45	AH	UK/ETO (Norwegian)
335	5	1944-46	-	ME (Greek)
336	5	1943-46	-	ME (Greek)
340	2/5/9/16	1941-45	GW	UK/ETO (Free French)
341	5/9/16	1943-45	NL	UK/ETO (Free French)
345	5/9/16	1944-45	2Y	UK/ETO (Free French)
349	5/9/16	1943-46	GE	UK/ETO (Belgian)
350	2/5/9/14/16	1941-46	MN	UK/ETO (Belgian)
352	5	1944-45	-	ME (Yugoslav)
400	11	1944-45	-	ETO, photo-reconnaissance (RCAF)
401	2/5/9/14/16	1941-45	YO	UK/ETO (RCAF)
402	5/9/14/16	1942-45	AE	UK/ETO (RCAF)
403	1/2/5/9/16	1941-45	KH	UK/ETO (RCAF)
411	1/2/5/9/14/16	1941-46	DB	UK/ETO (RCAF)
412	2/5/9/14/16	1941-46	VZ	UK/ETO (RCAF)
414	9/12/14	1944-45	-	UK/ETO, fighter-reconnaissance (RCAF)
416	2/5/9/14/16	1941-46	DN	UK/ETO (RCAF)
417	5/8/9	1942-45	AN	ME (RCAF)
421	5/9/16	1942-45	AU	UK/ETO (RCAF)
430	12/14	1944-45	G9	ETO, fighter-reconnaissance (RCAF)
441	5/9	1944	9G	ETO (RCAF)
442	5/9	1944	Y2	ETO (RCAF)
443	5/9/14/16	1944-46	2I	ETO (RCAF)
451	5/9/14	1943-46	BQ/NI	ME/ETO (RAAF)
452	1/2/5/8	1941-45	UD/QY	UK/Aus/FE (RAAF)
453	5/9/14/16	1942-45	FU	UK/ETO (RAAF)
457	2/5/8	1941-45	BP/XB/ZP	UK/Aus/FE (RAAF)
485	1/2/5/9/16	1941-45	OU	UK/ETO (RNZAF)
501	1/2/5/9	1941-45	SD	UK
	16	1946-49	RAB	UK
502	22	1948-51	RAC/V9	UK
504	2/5/6/9	1941-45	TM	UK
	22	1948-49	RAD/TM	UK
519	6/7	1943-45	Z9	UK, meteorological
520	5	1944	2M	ME, meteorological
521	5	1942-43	5O	UK, meteorological
541	4/5/9/10/11/19	1942-51*	-	UK, photo-reconnaissance (* Lancasters in 1946)
542	4/7/9/10/11/19	1942-45	-	UK, photo-reconnaissance
543	4/5/11	1942-43	-	UK, photo-reconnaissance
544	4/9/11	1942-43	-	UK, photo-reconnaissance
548	8	1944-45	TS	Aust
549	8	1944-45	ZF	Aust
567	5/16	1945-46	I4	UK, anti aircraft co-operation
577	5/16	1945-46	3Y	UK, anti aircraft co-operation
587	16	1945-46	M4	UK, anti aircraft co-operation
595	5/9/12/16/21	1944-49	7B	UK, anti aircraft co-operation
600	14/21/22	1946-50	RAG	UK
601	5/8/9/16	1942-49	UF/RAH	ME/UK
602	1/2/5/6/9/16/ 14/21/22	1939-51	ZT/LO/RAI	UK/ETO/UK
603	1/2/5	1939-42*	RL/XT/T	UK/ME (* flew Beaufighters 1943-44)
	16/22	1945-51	9R/RAJ/XT	
604	16	1946-50	NG/RAK	UK
607	5/8/14/22	1943-51	AF/RAN/LA	FE/UK
608	22	1948-51	RAQ/6T	UK
609	1/2/5	1939-42	BL/PR	UK
	16	1948-51	RAP/PR	UK
610	1/2/5/14/22	1939-51	JE/DW/RAQ	UK/ETO/UK
611	1/2/5/7/9/14/22	1939-51*	FY/RAR	UK (* flew Mustangs 1945)
612	16/14	1946-51	RAS/8W	UK
613	14/22	1947-51	RAT/Q3	UK
614	16/22	1946-50	RAU/7A	UK
615	5/8/14/21/22	1943-50*	KW/RAV/V6	FE/UK (* disbanded 9/45-10/46)
616	1/2/5/6/7	1939-44	QJ/YQ	UK
631	16	1945-49	6D	UK, anti aircraft co-operation
667	16	1945	U4	UK, anti aircraft co-operation
680	4/5/6/11	1943-46	-	ME, photo-reconnaissance
681	4/11/19	1943-46	-	FE, photo-reconnaissance
682	4/11/19	1943-45	-	ME, photo-reconnaissance
683	4/11/19	1943-45	-	ME, photo-reconnaissance
691	16	1945-49	5S	UK, anti aircraft co-operation
695	5/16	1944-49	4M/8Q	UK, anti aircraft co-operation
1435	5/9	1942-45	V	ME (from 1435 Flight, Malta)

RN FAA SEAFIRE SQUADRONS

Note: This table uses similar abbreviations to the 'RAF Spitfire Squadrons' listing. The columns are slightly different in that there are no squadron codes (due to the many duplications and variations) and the aircraft carriers from which the units flew are listed.

Sqdn	Mark/s	Dates	Operations/Carriers
800	15/17/47	1946-50	UK/ME/FE; Triumph
801	1/2/3/15/17	1942-47	UK/ME/FE/UK; Furious, Implacable
802	15/17	1945-47	UK; Premier, Vengeance, Venerable
803	15	1945-47	UK, Canada; Warrior
804	15/47	1946-49	UK; Theseus, Ocean, Glory
805	3/15/17	1945-48	FE/ME/UK; Ocean
806	15	1945-46	UK; Glory
807	1/2/3/17	1942-47	UK/ME/FE/UK; Furious, Indomitable, Battler, Hunter, Implacable, Vengeance
808	1/2/3	1942-44	UK/ME/UK; Battler, Hunter
809	2/3/17	1943-46	UK/ME/FE/UK; Unicorn, Stalker
816	2	1943	UK; Tracker
833	2/3	1943	UK/ME; Stalker
834	2/3	1943-44	UK/ME/FE/UK; Hunter, Battler
842	1/2	1943	UK; Fencer
879	2/3	1943-45	UK/ME; Attacker
880	1/2	1942-44	UK/ME/FE; Argus, Indomitable, Stalker, Furious, Implacable

Sqdn	Mark/s	Dates	Operations/Carriers
883	15	1947-48	Canada (shore)
884	2	1942-43	UK; Victorious, Argus
885	1/2/3	1942-44	UK/ME/UK; Formidable
886	2/3	1943-44	UK; Attacker
887	1/2/3	1942-46	UK/ME/FE; Unicorn, Indefatigable, Implacable
889	3	1944-45	FE; Atheling, Indomitable
894	2/3	1943-46	UK/ME/FE/UK; Illustrious, Unicorn, Implacable, Indefatigable
897	2/3	1943-44	UK/ME/UK; Unicorn, Stalker
899	2/3	1943-45	UK/ME; Indomitable, Hunter, Khedive, Chaser

Reserve squadrons: 1831 (F.17 1947-51); 1832 (F/FR.17/FR.46 1947-51); 1833 (F.17/FR.47 1947-53); all usually shore based in UK.

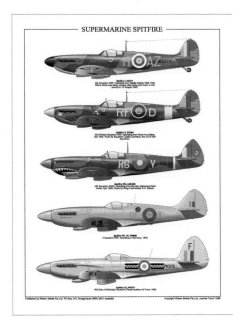